Integrative Design

Ralf Michel
(Ed.)

Integrative Design

Essays and Projects on Design Research

Birkhäuser
Basel

CONTENTS

PREFACE

Ralf Michel

Integrative Design: The Outlines of a Concept

Almost half a century ago, the Club of Rome published its first report on 'The Limits to Growth' (Meadows et al. 1972). Since then, world population has more than doubled, from 3.5 billion to more than 7.5 billion. Late in 2018, the German astronaut Alexander Gerst returned to Earth after more than six months on the International Space Station. Shortly before his return flight, he sent a video message to his yet unborn grandchildren[1] apologising for his generation: 'At the moment, it looks like we, my generation, are not going to be leaving the planet in the best condition'. Humanity is disrupting the climate, clearing forests, polluting the oceans, and we are consuming limited resources far too quickly. The Earth is a 'fragile spaceship', and he hopes that 'we can still get our act together'. On his return, he said that as a child, he dreamed of flying into space, believing it to be the most extraordinary place. After more than six months in space, however, he realised that Earth itself was the most extraordinary place in the universe – and moreover a very delicate one. Of course, Richard Buckminster Fuller and others recognised this long before Alexander Gerst. But almost no one has expressed the vulnerability of our planet so clearly, understandably, and emotionally to the general public, and at the same time placed our own responsibility at the centre.

But what does all of this have to do with design? The challenges Alexander Gerst saw so strikingly from outer space call for action on everyone's part. Not some day, not tomorrow, but today. Designers need to be aware of the implications of their actions, and must ask how their proposals lead to decisions or improve situations. According to Jörg Petruschat,[2] design generates power, and power has to do with the possibilities designers offer others through their designs.

Design reflects society and its developments in relation to issues that are addressed by the protagonists and thought leaders of the design disciplines, especially design researchers. In the future, a key term for the positioning of designers, the editor believes, may well be integrative design. This refers to the potential of an approach to design that is rooted in criticism, is intrinsically linked to political and economic positions, and that involves itself in sovereign ways in transforming the world based on a radically sustainable attitude. Or, as Tomás Maldonado proclaimed in 2009: 'The path we have to pursue is an open, critical and attentive rationality, especially against the grave problems of our time'. A world in which the challenges are not only complex, but in which the very complexity of the problems is complicated, and in which, from the outset, solutions are under increasing pressure to really make the world a better place.

The common theme of this book is the idea that designers, as the protagonists of a discrete and often misunderstood discipline, need to redefine their role in collaboration with technicians, economists, and politicians, but mainly in their relationship with the key stakeholders: those who attach importance to their artefacts.[3] From time to time, the paradigmatic conditions under which these roles are granted to them undergo change. Penetrated by the rationality of modernity, design has gone beyond the turmoil of postmodernism and moved towards blurred and not very productive concepts of the creative. Always bearing in mind the misunderstanding that creative action is somehow inexplicable, individual and, at best, ingenious. From recognition as an elusive economic factor to pride in authorship, designers pass through the shoals of superficial styling, which trap us with a consumerist level of consciousness. Today, so-called 'design thinking' is practised at management seminars. Creative design is misconstrued as an empirical method of variant formation. Design becomes recipe. Now, it resembles cooking: through experience, good chefs have internalised a sense of materiality and a taste for ingredients and types of preparation. They combine, vary, and create new connections based on this inherent knowledge. New roles in design are not created by simplification, and all attempts to generate recipes fail, because they suggest shortcuts.

This book attempts to describe the role of design in the culture of its integrative possibilities. It is not about a new design method, but instead about becoming conscious and about communicating. The point is to acknowledge, as a designer, and in all seriousness, that many people are part of the realisation of new possibilities and solutions, and that the role of the designer is to develop and visualise these possibilities and solutions in a sensual, meaningful, physical, and tangible way. The book *Integrative Design* brings together fundamental essays on aspects of integrative design; the associated website[4] documents recently completed research projects that address these aspects.

Jörg Petruschat (p. 11) explores the essence of design and calls for the independence/autonomy of design – and thereby moves into a controversial dimension that marks the limits of intercultural understanding between English and German-speaking audiences, which is why original language concepts are used at times in the translation. The term design, for instance, is followed by *(gestalten/entwerfen)*. Arguing historically, Petruschat's contribution goes in search of the autonomy of design and reflects on processes of consciousness raising and the heightening of awareness. Finally, his argumentation leads towards the self-image and, yes, the attitude of the designer. The discipline consists in showing possibilities to those for whom they design. And so, in his 'personal etymology', he connects doing *(machen)* with liking *(mögen)*: the designer as humanist.

Cameron Tonkinwise (p. 44) reflects on the future of design and design research in a dawning era of design after ownership. As Petruschat demands, he outlines the role of designers as thought leaders for an era in which not property but communally used artefacts are at the centre of the designer's interest. The post-industrialist design researcher must now focus on social practices and the ways in which new kinds of sociality are afforded by them. And Tonkinwise ends by stating that designers must become more politicised.

Tony Fry (p. 32) elaborates an unprecedentedly political agenda for design. He demands that designers reorient their positions to become truly future-oriented and political, '[...] which implies design becoming more dynamic, more powerful and more able to communicate the significance of designers to society in general. This means that the way designers think the culture they create and the practice they establish have to break radically with existing and dominant patterns.'[5]

The remaining five essays focus more specifically on the new artefacts with which designers and design researchers empower themselves through their competencies. Inclusion, social innovation, the role of direct creative action in the innovation process, and radical design in the context of participation and sustainability are such fields of action.

The role of design in the face of the challenge of inclusion was addressed by Tom Bieling (p. 97) in his dissertation, and now in condensed form as an essay in this book. The contributions of Sandra Groll (p. 113), who illuminates design as an interface with society, and Helge Oder (p. 128), who discovers future potential in the integration of design and technology, also draw upon dissertations. Ecological and social innovations as trendsetting basic positions in design, finally, are developed by Anna Meroni (p. 76) and Ursula Tischner (p. 57).

Without exception, all of these topics correspond to the UN's sustainability goals. However we may have heard or read about them, they mark the need for collective action on the part of our generation. They mark those fields of activity in which designers must take unambiguous positions. Political in the best sense, humanist in the best sense. Willing to compromise when it comes to solution strategies; uncompromising on core values. As a guiding concept, integrative design outlines a perspective for design that is genuinely involved. In addition to courage, its basis is practical reason, the ability to criticise and, of course, passion.

I thank the editor Nora Kempkens for her patience as we worked together to develop this book, Ian Pepper for his translations, and Sven Schrape for the graphic realisation. Thanks to Nicolas Ebner for his cooperation, and for his constant support, which accompanied the project right to the end.

And of course, I thank the authors for contributing to this initial approach to conceptualising integrative design; I hope we will be engaging in an ongoing discussion to sharpen our ideas. Specifically, we are building an internet platform where research projects that address the issues of integrative design can be published. Located at the following link will be an extension of this book in the form of concrete design research projects: https://www.masterstudiodesign.ch/publications/integrative-design.

Featured projects:

Eine Gretchenfrage an die Designtheorie
Oliver Baron
2012–13
Cape Peninsula University of Technology

Health Hardware Design
Christian Tietz
2012
Human Rights Award 2012
University of Technology Sydney

Lorm Hand – Communication Devices for deaf-blind People
Tom Bieling, Tiago Martins, Ulrike Gollner, Gesche Joost
2017–ongoing
Design Research Lab / Berlin University of the Arts

Hydrofix
Helge Oder
2015–17
HTW Dresden/ITU Dresden

Design for One World
Manuel Wüst, Patrick Müller, Ralf Michel
2017–ongoing
www.designforone.world / Institute Integrative Design – Masterstudio

1   Message to my future grandchildren, from Alexander Gerst, Commander ISS, 19 December 2018, ESA
    European Space Association, https://www.youtube.com/watch?v=4UfpkRFPIJk.
2   '"Wicked Problems": A Few Remarks on Design as Research' in this publication.
3   This position refers to the semantic turn, a theory of design that demands a paradigmatic change of
    perspective on the part of designers. The design of artefacts should primarily address their potential
    relevance to stakeholders rather than their sales-promoting rhetoric (Krippendorff 2005).
4   https://www.masterstudiodesign.ch/publications/integrative-design.
5   See 'An Unfolding Political Agenda' in this publication.

References

Krippendorff, K. (2005). *The Semantic Turn – A New Foundation for Design*. London: Taylor & Francis.
Meadows, D. L., Meadows, D. H., Randers, J., Behrens III, William W. (1972). *The Limits to Growth*. New York:
    Universe Books.

'WICKED PROBLEMS':
A FEW REMARKS ON DESIGN AS RESEARCH

Jörg Petruschat

At first glance, it's about money. In this society, it's always about money. Whoever poses the question of whether design can be regarded *as* research has an interest in seeing *that* design is regarded as research. Whoever asks this poses a question about legitimacy. The legitimation of design as research vies for recognition in two domains: in economics and in academia.

In the domain of economics, it is difficult for designers to label the peculiarity and irreplaceability of their activities. The space of explanation is already occupied: product development – and this is not only true in Germany – is regarded as the business of implementing concepts within material processes. And then, from time immemorial, there have been engineers, managers, and marketing specialists, and standing above these actors, powerful decision-makers. They arrive at decisions according to corporate interests – as everyone knows, it is not a question of chairs or blenders, but instead of money. When in doubt, therefore, the decision-makers orient themselves through the Excel spreadsheet of the controllers. They want to know the figures for sales volumes, revenues, and earnings. Because valorisation is a function of the velocity of circulation, the controllers, the decision-makers, but above all the others, those who pay for product development or are paid for it, want to know the value of their activity per unit of time.

What do designers do with their time? They design. But what is that? Generally speaking, to design something means to toss out ideas – is a relationship that is palpable in German linguistically, where to design is to *entwerfen*, where a throw is a *Wurf*, and to toss something out is *hinauswerfen*. Designers toss out their ideas in a very special way. They incorporate them in drawings, into virtual and physical models. That is their performance. Afterwards, the engineers arrive and turn this theatre into something that is usable for business. They too engage in design activity. They too toss out ideas. But they do so, obviously, in a different way. Thanks to the designers, they already have a concept in front of them. The engineers, as they are fond of saying, change existing situations into preferred ones. And because standing at the end of their efforts are well-dimensioned plans or formulae across which the eye can rove gladly, engineers too are the creators of what is referred to throughout the English-speaking world as 'design'.

Conditions for engineers are far tougher than those for designers. But precisely this physical, chemical, biological, and physiological toughness when it comes to the conditions of design faced by engineers actually makes design work far easier for them. They simply do what these conditions permit. Their ideas do not go beyond these constraints, but instead into them.

The design activity of engineers is subordinated to the task of adapting to the acknowledged rules of technical reality and of technology. This is required by the market, and of assurances provided to customers. Here, there are standards and agreed-upon procedures. Engineers must be familiar with these routines. And they are. It is what they were taught. A proficiency with these routines is their knowledge. The engineer is – as a product of education – a conservative and cunning creature. As soon as engineers depart from agreed-upon procedures, depart from routines, their work becomes risky too. They are held liable for calamities. When they venture to engage in work beyond agreed-upon procedures and routines, they call it research. It requires some leeway, and some space. They call these spaces laboratories so that everyone will understand that the research taking place in them is also a form of work. In short, engineers legitimise their researches not through that which they have to do anyway, but because the old routines for the implementation of ideas and material processes have become inadequate. Designers, meanwhile, are remote from such agreed-upon procedures.

The second domain within which designers strive after money, but also legitimation, is the academic enterprise.

Here, research is a constant. It is based on the developmental logic of the respective discipline. Recognised as research is that which follows the acknowledged rules of the academic enterprise, and which perpetuates a discipline's existence. To this extent, academic research too contains conservative elements. Within the academic enterprise, therefore, research that strives to establish new objects and methods also requires free space. The free spaces are situated between the disciplines, and are referred to as interdisciplinary or transdisciplinary research enterprises, and if laboratories also exist for this purpose, it is possible too to work on these free spaces themselves, so that new fields emerge in the realms between the disciplines.

In this context, designers have barely gained a foothold. Although they have attempted for decades to ennoble design via the supplement 'scientific', they are at best tolerated within the scientific fields. When it comes to sponsored research in Germany, design simply does not appear; here, as in general in the EU documents pertaining to budgetary decisions, their activities are consistently regarded as service provision. Here, the designers themselves are not entirely free of guilt. For decades, hoping to become economic players, they shouted the slogan: 'We are service providers!' at anyone who would listen. For ten years or so, the wording has been different. Now, designers are creative producers. They create ideas. Once again, they are playing the old cards of provocation and intervention, the card of art. Once again, they are original. That way, financing seems more assured. You pay artists, without asking too many questions, since, after all, they make art. These budgets too may be less generous in the future, but the status of art – with regard to the academic training of artists as well – remains undisputed. An academy of art and design has already renamed itself an art academy, striking the word 'design' from its letterhead. This may be due to the fact that ever fewer students want to study design.

In the context of the art academy, design can only legitimate itself as applied art, as art that makes itself useful. Here too, it is true for design: please stand at the end of the queue, we'll see what's left over for you from art, or: why don't you show us your usefulness as artists? Why, I am asking, did designers unequivocally want to exit this nexus sixty years ago?

In these preliminary remarks, I want to call attention again to a third attempt to legitimate design as research within the academic enterprise. This attempt consists in the subtle detection of the way the activity of designers has been a component of scientific activity for a long time anyway. Zealously, all of the sensuously graspable material that scientists have promoted in research processes – every notation, every sketch, every diagram, every model, everything that goes beyond alphanumeric formulations – is now recorded, discussed, referenced, and furnished with the label design. For many scholars in the humanities and social sciences, a lot of theories and discourses seem less metaphysical, less idealistic, less abstract in conjunction with these graspable, palpable objects, and hence more valid, richer, and a bit more 'objective' than all of the chatter about objects and methods could ever hope to have been.

But presumably, even this excursion into the cultures of experimentation will not grant design a ticket of admission to the concert of the academic disciplines. Not even when, after a 'linguistic turn' and an 'iconic turn', a 'design turn' is now proclaimed. For what is meant here with this recent turn is not design as an independent epistemic practice, but instead the availability of scientific knowledge (Schäffner 2010).[1] To the assertion that design is an applied art, the idea of a 'design turn' responds with the argument that design is applied science. But for this purpose, for the application of science, we already have engineers. A vicious cycle ensues. Suspended between art and science, design runs back and forth in search of a legitimation for its cognitive work, while resounding from the saturated disciplines are the words: We already have them.

In my words, the alternatives are: pander to or break away? Designers and their theoreticians have pandered since the 1950s. Not to the arts, from which they derive their origins and from which they have attempted, with good reason, to free themselves, but instead to the scientific and engineering disciplines.

I therefore want to develop a number of arguments that would make a break possible.[2]

Wicked Problems (and Nebulous Methods)

All research endeavours about, through, as, or towards design (and no matter how this set of references is prolonged or differentiated), all of these efforts find their ideal point of reference in the notion that design is concerned with solving problems. Design counts as a problem-solving discipline. We find, however, that the consensus concerning this definition is stronger than our understanding of what is meant by the term 'problem-solving'. In the very recent past, for example, Nigel Cross regarded design as being comparable to chess, and believed one could best investigate how design functions as praxis by watching the great masters over their shoulders.[3] Is this a pragmatic attempt to learn by imitation? Or does he still stand in the tradition of that rationalist school which believes that design can be fully demystified through the representation of methods? I have no wish to dispute this intention. But when a theory, with all of its epistemic equipment, can only demonstrate that designers solve their problems in the same manner as engineers or scientists, it remains incapable of highlighting the specificity of their occupation. It is hardly sufficient to postulate 'designerly ways of knowing, thinking, and acting' – they must be demonstrated with reference to the real activity of the designer. The criteria of specificity for 'designerly ways' cited by Cross are true for all creative work, and have not gone very far beyond the formulation of Herbert A. Simon, according to which 'engineers are not the only professional designers' (Simon 1996, 111).[4] The customary equation in English-language and thought between the terms *Entwurf* and 'design' gets in the way of a more precise definition of design as professional praxis. In the German word 'Entwurf', activities are addressed in very general terms through which intuitions or 'ideas' are 'thrown out' into realities, lead out into reality. What I mean here is that without a reasonably well-articulated theory of the particularity of sensuously founded epistemological models, without a reasonably framed hypothesis concerning the nature of the formation of a gestalt/design *(Gestaltwerdungen),* and of the way in which it is related to specific human capacities, explanatory models on design draw – often counter to their intentions – upon a romantic conception of genius that is ill-suited to the collaborative character of contemporary design processes and the status of the designer within them.

This begins already with the implicit acceptance of an idea from Herbert A. Simon, according to which designers conceive preferred states, which are then transformed into existing ones. But is this the case? Does the motivation for change emerge from a nebulous future? In my view, a problem emerges when routines malfunction. Here, I follow Karl Popper, who perceives things in a similar way.[5]

Problems arise when the routines in which all living beings reproduce their existence lead to threats to that existence, or even remain unsuccessful. Then, alternatives must be found. This however involves no special performance on the part of designers. Any crow can do that. Creativity, the efforts of the living creature to discover behavioural models that secure its existence, arises from the game of coping with disappointments and frustrations.[6] I will be returning to this topic soon.

One strategy for arguing for the singularity of design is offered by Richard Buchanan.

Somewhat laboriously, he disentangles himself from the epistemological model proposed by Simon and Newell (1972) in *Human Problem Solving*[7] and takes up a debate which arose in 1974 at the Design Theory Congress (Spillers 1974):[8] the debate about the viability of Horst Rittel's idea of 'wicked problems' (Buchanan 1992).[9]

The mathematician Horst Rittel, who developed a logically grounded theory of design *(Entwurfstheorie)* during the 1950s and 1960s, and who taught at the Hochschule für Gestaltung in Ulm, encountered a category of problems in the context of his work in the area of urban sociology and planning which defied the linear programming of planning processes (Rittel and Webber 1973).[10] It was a question of problems whose factors were in some sense messy, or in any event lacking in clarity, and hence not amenable to being transformed into manageable, logically structured tasks in the way familiar to engineers, the kind of problems that can be worked through ingeniously according to the formula of tables of calculations. Those involved in urban planning cannot simply project typologies of traffic flows and the resources for delivering supplies and for waste disposal according to engineering principles, but must also cope with a network of social factors, each of which displays an incalculable dynamism and complexity. The interaction of logically calculable and logically unpredictable factors and their ambiguous interrelationships makes the work perplexing.[11] A number of people who heard Rittel's lecture in New York came upon the idea of reclaiming the perplexing, vague qualities encountered in the initial conditions of problem solving as the specific challenge for design.

In the present context, it is important to recognise to begin with that responsible for the 'wickedness' identified by Horst Rittel with regard to complex problem areas are the limitations of a systematic planning theory. The term 'wicked problems' refers to the limits of what can be mastered by processes that are ramified in a linear way.[12] The model of linear operations, which allows computers to hum so promisingly, is no match for the complexity of everyday design praxis.

Although Richard Buchanan thematised certain issues with regard to the so-called 'wicked problems', the results of his efforts with regard to what actually designers do are disillusioning. In a typically Anglo-Saxon manner, Buchanan does not differentiate between the terms 'Entwurf' (i.e. a general sense of mental models that are thrown out into reality and thus bound back to reality) and 'design', making it difficult for me to discern what he regards as specific to professional design work. Finally, he perceives the perplexing quality of design problems in the fact that they are indeterminate (in contrast to being undeterminate/under-determinate). According to Buchanan, the special trait of designers is that they begin work with an explanation and specification of the object.[13] Finally, he argues, they work on something that does not yet exist. This, he says, differentiates them from scientists, who know from the beginning what they are doing because from the start, they act within regulated systems and in a sense only enhance their definiteness.[14] This proposal for difference goes one step further – it shows that design is obviously something other than science, and that this something other has something to do with processes of clarification that function differently from scientific methods.

Nevertheless I find Buchanan's proposal ill-suited to enhancing the reputation of the designer in concert with research, since it amounts to saying that: Here are people who, at the start, never know exactly where they are going with the things they are doing. I do not deny that such uncertainty surfaces repeatedly, but it can hardly be declared to be the core of professional competence. Isn't it a question of instead demonstrating how designers succeed in overcoming general uncertainties? Isn't it necessary to show in detail how designers succeed in taming wickedness, which is to say proliferating complexity?

In opposition to Buchanan, I do not perceive the specificity of design problems in there being too little information on the table. Design does not begin in vagueness, like a spirit that hovers above the waters. The design conception always begins with a reality – that is the case. The primary art of design however consists in calling this reality into question, in dissolving the pre-existing forms and their order, which has up to now appeared to be compulsory.[15] The vague and indefinite aspect of the design problematic is a self-created drama, not a special fate that clings to it.[16]

Design begins with the recognition that the factors on the table are no longer compatible with one another. More precisely: design begins with a recognition that the arbitrariness that has succeeded up to now in bringing the effective factors formally into a whole is no longer acceptable.[17]

Design begins with a critique and a disorganisation of reality, with the liberation and redemption of forms and functional models from their previous dispositions and contexts of application, with the destruction of a reality that is imagined as an integrated and functioning whole in the object.[18] Designers dissolve – initially for themselves – the hitherto familiar performance of use and enjoyment. At least initially, then, this work on the dissolution of an existing order produces that vagueness which Buchanan identifies, not without justice, as a particular feature of design processes.[19]

It therefore falls short to see the special nature of design work solely in a trajectory from the vague to the formally determinate. As a praxis of professionals, design to begin with ascertains what is confounded in this world. Who, if not designers, are going to expose the beautiful appearance of the regulated world as deception and experience it as repellent?

They begin, then, by transforming the existing, the fixed, into something vague and virtual. *That is the problem*. They are competent, first and foremost, to destroy the surfaces and arrangements through which things and experiences have hitherto cohered. They achieve this, however, only by demolishing the aura of success, depriving the object of its formal wholeness, and dissolving the structure of the components that remains.

I am arguing for a radical reformulation of our notions about the design process: the first step in design is not the idea, not inspiration. Whoever positions an idea at the inception of design processes only prolongs Platonic idealism. The creative process – and this is finally getting talked about – always begins with a confrontation with material things, with the insufficiency of habit and routine. This confrontation, a refusal of the absence of feeling, of thoughtlessness, in interactive relationships also leads towards a conscious analysis of how the markers of success and failure reach deep into the unconscious. An idea does not enter the brain like a bolt of lightning. An idea emerges from the ground of all experience, and with the setting into motion of our bodies via the drawing surface, and through model building, we can promote its emergence. It is a rhythm, a grammar of our physiology which endows the non-conscious data cluster in our central nervous systems with a recognisable structure and incisiveness. There is no form-idea somewhere above us, in which we might partake. An idea is nothing but the integration of experiences that are propelled into consciousness. And the driver for this consists of the design tools and materials that are set into motion physically.

In a historical period when it is a question of a red wine goblet and wallpaper patterns, the notion of an inner idea as the origin of its external realisation might have still sufficed. In today's complex reality, design processes begin with the hacking of culture and its hardware. The elements or factors of the problems are themselves complex entities. Which of these elements is still usable? Which must be changed? How are the hierarchies between them to be reordered?

Concerning these components of a complex context, designers often cannot know in what their effectivity is grounded in detail. Currently, the materials that come into their hands are no longer dumb like plaster and foam, although – for the sake of performance – they often elaborate visual models using them. These are smart materials, electrical components, microcontrollers, which are attached to other intelligent networks. The game that unfolds today in the design is neither created nor calculable beforehand – it is athletic.

As we see, the thesis that design works with certain sorts of problem, the ones that are called 'wicked', remains on the table. On the contrary: I am advocating getting to the bottom of this 'wickedness'.[20]

In my opinion, success is more likely when one avoids perceiving 'wickedness' in the complexity of problems. Such problems appeared 'wicked' to Horst Rittel because he sought to bring the planning process in line with the paradigm of artificial intelligence. But human actors do not behave in a causal and linear way through time. They are themselves complex beings. The complexity that presents itself to them in sensory form represents a challenge to adopt a position in relation to it. (This connection seems intuitive in German, where the word *Gestaltung* [design] is derived from 'Gestellt-Sein' [to be positioned], so that *Gestaltung*/design means the taking of a position.) *(Gestaltung ist Stellungnahme. Vom Gestellt-Sein kommt das Wort Gestalt her.)* That the models of rationality fostered by computer models fail when confronted with 'wickedness' does not mean that no other approaches to processing complexity are up to the task.[21]

I will explain here briefly.

With a new form, undoubtedly, designers ought to shed more light on the affairs of this world. And in doing so, doubtless, they move along a trajectory from a preconscious or no-longer-conscious state towards a consciousness of a world that can be experienced in sensory terms. I myself have described this often enough.[22] *Gestaltung* means becoming conscious. But, as Sergei Eisenstein has said, 'To recognise is to construct' (2006, 81). And construction is something different from mere intellectual enlightenment.

Design presents individuals with objects, that is to say conditions, that have been socially constructed. The job of the designer consists in bringing the multiplicity and ambiguity of technical, economic, social, cultural, sustainability-related, and other factors into a specific, finite, complete form. Designers are the agents of wholeness. In the process, they use the psycho-physiological mechanisms of structural formation that have developed within them through the retina, the ears, the hands, all the way to the brain.[23]

As agents of wholeness, as workers on the composing of forms *(Gestalten)*, they must also, however, be masters of framing.[24]

If designers are to succeed in subsuming[25] complex structural factors into a single gestalt, freely floating cascades of complexity must be halted.[26] This demarcation of boundaries along the margins of wholeness is no easy task, and does not come about spontaneously. Here, I don't want to give the impression that cascading complexity is only a 'wicked' problem for routines that are organised in a linear fashion.

The process of gestalt formation too encounters processing boundaries, but in contradistinction to the logical processing of causal chains, processes of gestalt formation are indebted not just to causality, but also, and first and foremost, to the principle of selection.[27]

Selection means the formation of hierarchies.[28] Gestalt formation not only organises the juxtaposition of the elements, but also delimits foreground from background, establishes priorities.

It is one of Herbert Simon's lasting achievements to have developed a theory of complexity which organises complexity into hierarchical levels. Often overlooked, however, is the fact that this theory of complexity is not based on the way in which a computer functions, but instead on an analysis of psychological experiments on human capacities for comprehension. Admittedly, Simon only regards human intelligence as the prototype of artificial intelligence. In contrast, I find it more interesting to derive knowledge and usefulness from the difference and intrinsic value of these two forms of intelligence. I can only deal here critically with Herbert Simon, but I do want to draw attention to the fact that Simon introduced a profoundly subjective category for the termination of cascading complexity, that of satisfaction. From the perspective of artificial intelligence, the recognition of the deeply human criterion of satisfaction is a sign of resignation. According to Herbert Simon, 'No one in his right mind will satisfice if he can equally well optimize; no one will settle for good or better if he can have the best' (1996, 119). This is the language of efficient calculation. What we, however, require for immediately comprehensible reasons – and I refer here to the debates on the limits of growth – is not mere efficiency, but above all sufficiency. In my view, the category of satisfaction is not a stopgap, because parallel processing is too slow, but instead an existentially vital measure. Only the power of boundary demarcation is a help against the proliferation of complexity, since otherwise, the whole always diverges uncontrollably. The decision for this interruption is at the same time the decision for a configuration of factors and their positive arrangement.[29]

Culture

The strength and the aesthetic specificity of design as an occupation consists in grasping unregulated but also frustrating situations intuitively and holistically; grasping means to pattern, and hence to order.[30]

Such patterning and decisions concerning framing are based on experiences which reside only to some extent in the specific factors involved. For the formal grasping of factors that do not stand in a logical relationship to one another, designers bring into play embodied and often only unconsciously present aesthetic and cultural experiences.[31] I refer to this ordering work here as information, as sensory knowledge. By giving a cluster of data blocks a framing, designers produce a difference with the surrounding space; by giving a data cluster a specific structure, they enrich the constellations semantically. Emerging in this instituted order is information that previously, in the mere juxtaposition of factors, was not the case, was not yet present.

What designers do here, in popular speech, is to formulate the realisation that further increases in complexity are not satisfactory [against the background of their cultural experiences].[32] Less – and many are fond of the citation – is more. But the question is: What is satisfaction? I will speak soon about this in more detail. First, I want to refer to an analogy: with regard to many of its processual traits, design activity is very similar to the production of consciousness in the individual. Consciousness, the generation of the human individual, the notion of reality and the specific knowledge that it is only your 'film', which runs in your imagination, this process too is based upon a degeneration of complex quantities of data. Consciousness is not somehow added to the events which occur in the central nervous system.

Consciousness is a selection from all of the data that converges at every millisecond in the central nervous system from all of the areas of the body. When we see, when we experience ourselves in sensory terms, we assemble only a highly restricted selection from this data to form a scene that enables us to act and to make decisions. And this selection, this order – like the act of aesthetic pattern formation – is also the generation of information on the basis of rejected complexity.[33] What we can learn from this is that the framing of complexity is not a task, not resignation, but the generation of a new quality.

I had to lead up to this point, because it is only now that I can show clearly that decisions to interrupt the growth of complexity are based on contexts that are more deeply anchored than any current problem definition. In brief, I refer to this context of satisfaction that relativises all current problems as 'culture'.[34]

You might answer by saying that it is old hat to say that design allows cultural competencies to flow into design activity. And I would have to reply: Yes. It may be old. But I am concerned with far more than that. In accounts of design activity that are familiar to me so far, culture is a factor that must be considered because it represents a medium into which the object must be inserted.[35]

Design objects in a culture, however, are something very different than fish in an aquarium. My proposal for researching the cultural factor in design is therefore more far-reaching. I do not see culture as one factor *in* design processes, not as a factor *alongside* others, but instead as the in the deepest sense decisive and formative *(gestaltbildenden)* one above all others. Here, I must therefore make a little detour into the arts.

From the artistic process, we learn that only a social recognition of aesthetically produced objects converts them into 'art'. We can speak of art only after a work has entered a process of social confrontation, after it has been elevated by the galleries, the art market, by reviews. Similarly, I would propose that the design process has not come to an end when the design object arrives on the market or reaches users. This only perpetuates the old commodity aesthetic, for which all 'designerly ways of knowing, thinking, and acting' terminate in the exchange of money for commodities. The problem for design does not actually consist in solving problems – although it does that as well. I have characterised the decisiveness that is summoned in this context to interrupt the accumulation of complexity. But precisely through interruption, this closure of a potentially interminable identification of the possibilities of production, of the distribution of actors and resources, of social embeddedness, and so forth, a position is presented in society that is not to be merely registered passively, but to be evaluated and recognised within aesthetic experience. Design products are proposals for grasping social complexity at a given moment and to a specific measure, and the experience of stability of this measure *(Maßhaltigkeit)* is the basis for their enjoyment. In that objects present experiences, something new is generated in the specific form through which a problem seems to be resolved, because emerging within the confrontation by the user with the object, in turn, is a decision situation. Here, it is not simply a question of: Do you want to buy me? But instead, quite simply: Do you agree with me? Do you resonate with the specific proposal, inherent in the object-form, which is presented here, for example concerning the interrelationship of resources and actors? These problem statements are not theoretical, nor are they accessible only to theory. They are present in sensory form, and refer back to experiences the user has lived through also in very different domains.[36] Regardless of whether resonance or dissonance is experienced in relation to the design object, in both cases an awareness emerges on the side of the user of the difference and autonomy between him/herself and those responsible for formal decisions regarding the becoming-form *(Gestaltwerdung)* of the product.[37]

Knowledge in Design and its Research

I would like to make another remark on this issue as my guideline for perceiving in design not a dimension of production, but instead a dimension of reproduction (with reference to the relationship between work and enjoyment).

In his famous formulation, Bruce Archer calls research a systematic questioning whose aim is knowledge. And with reference to Archer, Nigel Cross leaves no doubt that he regards this knowledge as an intellectual inventory, as a point of view that is oriented towards reflection. I would like to call attention to the fact that a knowledge that is reliable, that can be made available to others in a usable form, is not merely the symbolically coded knowledge of scientific languages, but is instead present in the sensory gestalt of objects or media contents. A research that is present in the results of materially conceived objects and practical processes is just as valuable to design as a research that ends in conceptual discourse.

True for both forms of research, for 'research from below', which emerges from the practice of design and its material and habitual conditions, as well as for 'research from above', carried out from the heights of reflection on aesthetic capacities and cultural involvement is: design is not fulfilled in the object or in the process; instead, design is the objectified instrumentality of the active, decision-making subject, i.e. between actors from the domain of production and actors from the domain of consumption.[38]

In so tracing design back to culture, from which it is decided and developed, and into which it returns, the term *Entwurf* (see above on the etymology of the term in German) acquires a deeper etymological significance for design as well. The German word *Entwurf* (design, plan, conception, scheme, outline, etc.) does not (as Vilém Flusser was able to convince many people a few years ago) have something to do only with the *Werfen* (pitching, tossing out) or *Hinauswerfen* (rejection) of ideas, inspirations, and so forth. The word *Ent-Wurf* means and refers to the reconnection of ideas and concepts to pragmatics. *Entwerfen* relates to *Werfen* exactly as

Enttäuschung (disillusion) relates to *Täuschung* (illusion), which is to say a reversed, enlightening, and critical attitude. Drafting and prototyping not only return high-flying ideas, which have become substantially loosened from reality, back to the ground of the facts, but first and foremost to the background of culture. This is why the roots of design lie in the designating of objects, in rhetorical categories such as 'ornare' and 'decorum'.[39]

If I argue here for expanding the domain of research of design so radically that it also thematises questions of the cultural development of a society, then this to begin with has a simple pragmatic reason: without a broad conception of the design process, without thematising in the cultural background, it becomes impossible to argue adequately in favour of cognitive achievements of design. What is special and indispensable about the capacity for *Gestaltung*? That forms are altered? When I board a streetcar and validate my ticket in a machine, this too is a change of form. We however concede that this act is also *Gestaltung* (design)?

I call *Gestaltung* the extension of life possibilities according to the requirements of a self-model.[40] I do so with the intention of distinguishing *Gestaltung* from near modifications of form, and to establish a connection with cultural progress.[41]

If design itself is conceived as a cognitive, which is to say a knowledge-creating discipline, as Cross demands, as a specific kind of knowledge, thought, and action, we must identify the capability which adds things, inputs to the culture that are more than cultural stylisations or performances by technicians and researchers from other disciplines. This is completely independent of my own deliberations, indicated here only briefly.

Required if we are to argue for such advances in knowledge is a general frame of reference that supplies and makes available the required criteria. And I call this frame of reference precisely 'culture'. Culture is not merely an aggregation of spiritual values, nor does it simply consist of a set of techniques that are re-inherited from generation to generation as generators of conformity. By culture I mean the totality of conditions that make it possible for the individual to make oneself special. I called culture the pressure for conformity that guarantees social reproduction in which to provide the individual with possibilities to become non-conforming and to generate the richness of individual differences. This also includes the intellectual and bodily preconditions, talents, competencies, and self-evidently the complex reality of material things within which these competencies and talents accumulate and are socially distributed. One of the central thematic areas regarding the development of culture through design is the question: Which degree of complexity, and under which constellations, are individuals not only to understand, but also to implement in the formation of their own individualities and agencies? For whoever engages in design poses questions of power. In particular, design calls the power of others into question. If they are good at their job, they present us with performances of how the lives of others have been lived up to that point. Whoever stages such a theatre, whoever creates models in this virtuoso way, which others are expected to take for reality, and whoever sketches out other possibilities in these models, i.e. for the sake of marketing or living with greater happiness, arrogates to themselves the role of an individual or occupational group that is in the possession of knowledge about the happiness of others. This structure of guidelines and proposals necessarily establishes an imbalance of power – those who know how to make others happy lay claim to the interpretative authority over the lives of others over their capacity for enjoyment and their ways of being happy. They decide whether design merely simulates sovereignty, as Peter Sloterdijk cynically attests, or whether it actually offers individuals opportunities to grasp, to differentiate, and to develop their competencies in relation to the complex problems of the world. Are designers so to speak mere *Standartenführer* [the term refers to a rank within Nazi paramilitary organisations – transl.] of taste, as suggested by Mateo Kries, in an almost unsurpassable gesture of poor taste, when he addresses the question to the German people: 'Do you want total design?'[42] Or do they succeed in distributing and socialising their power? Power is not just overpowering; the concept of power also – at least in my personal etymology of the German language – relates to possibility, empowerment to enabling, making to liking.

These questions are never absolute, can never be answered only with reference to preferred situations, in line with Herbert Simons thinking. Which is why this model is so antiquated. These questions can only be answered against a tangible horizon and with much redundancy, since accumulated in the objects which condition all of our lives are experiences of the mastering of complexity in each current manifestation. This is the special aspect that confronts designers in their work, and to which they answer through their responses to the development or the simulation of competencies.

1   'Occurring currently in the natural sciences is a fundamental turn: the analysis of nature has arrived at a point at which the direction of research is reversed. Now, it is no longer a question of investigating the processes of nature, but instead of how humankind might proceed in a different direction using basic elements' (Schäffner 2010, 33). And: 'This turn toward "doing things", towards making, is the essence of this turn toward design, a "design turn" that is bound up directly with the nanotechnology revolution' (Schäffner 2010, 36). A similar diagnosis preoccupied Edmund Husserl earlier, and impelled him to write his late work *The Crisis of European Sciences and Transcendental Phenomenology*. The central conceptual figure in this work is the dialectic between image and method, between the 'what' and the 'how'. In theorising it, Husserl gives prominence to Galileo. It was Galileo who transformed geometry from a science of the description of origins to a formula that could be applied advantageously in the most diverse contexts. Since Galileo, the sciences have no longer evolved in relation to objective foundations, but instead in terms of method, and this has had consequences that penetrate all the way into the everyday lifeworld: instead of knowing how a doorbell rings, it is enough to know how to ring it. (See Hans Blumenberg's essay on Edmund Husserl in *Wirklichkeiten in denen wir leben*.) With the idea of a 'design turn', design – as an occupation and a professional praxis in contradistinction to design as an applied science in the sense of engineering – is declared to be an occurrence whose decisive turn still lies before it. 'While the natural sciences, along with engineering, have long since practised a turn toward design, it is not only the humanities and social sciences that stand before a decisive turn, but also the design disciplines.

Here, it is a question in particular of the replacement of the individual designer with the black box of his unconscious creativity through an interdisciplinary laboratory. Design strategies will be substantially modified through the experimentalisation and through a close association with analytical historical knowledge of the humanities and social sciences' (Schäffner 2010, 40). Here, Schäffner does not discuss the wealth of experiences professional designers can show for themselves since the end of the Second World War in multi-competent personnel arrangements with collective prototyping and in experimental contexts.

2   In all fairness, I must however say something more about my position. In the epistemic praxis of design, I see primarily a 'hands-on' praxis. I must say this clearly, because an epistemic practice can also mean brooding over books. Theory too has its praxis, and the turning of the pages of the book contains a manual element. The aim of theory however consists in converting reality into abstract models. The aim of a design model is to engender palpable realities using concrete models. While theory reduces complexity, design generates it. I don't know whether you will be satisfied at this point with this *bon mot*, since it could be objected with justice that every scientist who constructs prototypes with the engineer of his laboratory in order to assess the applicability of his theories, elevates complexity. To say nothing of artists. If I want to provide design with a positioning in free space that is distinct from science and art, a reference to the increase of complexity hardly suffices. Admittedly, the peculiarity of design as a praxis that produces knowledge cannot be grounded simply by pointing out that design is something different from theory.

3   Cross does not distinguish between rationalistic thought and design. He positions himself in the rationalistic tradition of Horst Rittel and speaks in the same breath of 'wicked problems' and of 'designerly ways of knowledge, thinking and acting' (Cross 2007, 44 ff).

4   This citation continues: 'Everyone designs who devises courses of action aimed at changing existing situations into preferred ones. The intellectual activity that produces material artifacts is no different fundamentally from the one that prescribes remedies for a sick patient or the one that devises a new sales plan for a company or a social welfare policy for a state. Design, so construed, is the core of all professional training: it is the principal mark that distinguishes the professions from the sciences. Schools of engineering, as well as schools of architecture, business, education, law, and medicine, are all centrally concerned with the process of design' (Simon 1996, 111).

5   In 1972, in his 'The Logic and Evolution of Scientific Theory', Popper argues for regarding science as a biological phenomenon, one that is rooted in pre-scientific knowledge: 'a problem arises for the animal when an expectation proves to have been wrong. This then leads to testing movements, to attempts to replace the wrong expectation with a new one. [...] The problem arises when some kind of disturbance takes place – a disturbance either of innate expectations or of expectations that have been discovered or learnt through trial and error' (Popper 1999, 4).

6   See Donald W. Winnicott's *Playing and Reality* (1971) and my own 2011 lecture 'Fassungslosigkeit. Einige Bemerkungen zum freien Spiel der Kräfte'.

7   'A person is confronted with a problem, when he wants something and does not know immediately what series of actions he can perform to get it' (Simon and Newell 1972, 72). And: 'To have a problem implies (at least) that certain information is given to the problem solver: information about what is desired, under what conditions, by means of what tools and operations, starting with what initial information and with access to what resources. The problem solver has an interpretation of this information – exactly that information which allows us to label part of it as *goal*, another part as *side conditions*, and so on' (Simon and Newell 1972, 73).
Elsewhere, we read: 'If a person is to solve a problem, there are several things he must know. First, he must know the set of problem elements – that is the materials of the problem. Second, he must know the initial state of the problem and its goal. Third, he must know an operator or a set of operators for transforming the initial state into the goal. Finally, he must know the restrictions under which the operator may be applied' (Hayes and Simon 1979, 169/170).

8   The conference, funded by the National Science Foundation, was held at Columbia University.

9   'There are so many examples of conceptual repositioning in design that it is surprising no one has recognized the systematic pattern of invention that lies behind design thinking in the twentieth century. The pattern is found not in a set of categories but in a rich, diverse, and changing set of placements, such as those identified by signs, things, actions, and thoughts. Understanding the difference between a category and a placement is essential if design thinking is to be regarded as more than a series of creative accidents. Categories have fixed meanings that are accepted within the framework of a theory or a philosophy, and serve as the basis for analyzing what already exists. Placements have boundaries to shape

and constrain meaning, but are not rigidly fixed and determinate. The boundary of a placement gives a context or orientation to thinking, but the application to a specific situation can generate a new perception of that situation and, hence, a new possibility to be tested. Therefore, placements are sources of new ideas and possibilities when applied to problems in concrete circumstances' (Buchanan 1992, 13). And Buchanan adds in a note: 'The concept of placements will remain difficult to grasp as long as individuals are trained to believe that the only path of reasoning begins with categories and proceeds in deductive chains of propositions. Designers are concerned with invention as well as judgment, and their reasoning is practical because it takes place in situations where the results are influenced by diverse opinions' (Buchanan 1992, 13).

10 'The kinds of problems that planners deal with – societal problems – are inherently different from the problems that scientists and perhaps some classes of engineers deal with. Planning problems are inherently wicked. As distinguished from problems in the natural sciences, which are definable and separable and may have solutions that are findable, the problems of governmental planning – and especially those of social or policy planning – are ill-defined; and they rely upon elusive political judgment for resolution. (Not "solution". Social problems are never solved. At best they are only re-solved – over and over again)' (Ritter and Webber 1973, 160).

11 Horst Rittel therefore saw linear planning systematics as experiencing a crisis, and developed a second order systems theory in order to overcome it. In this context, he conceded that actually, all planning problems are 'wicked' (Rittel 1992, 48). 'In general, one can say that the era of hope and expectation with regards to systematic approaches has been supplanted by an era of disappointment. Especially in the United States, we encounter a disenchantment regarding the possibilities and practicability of these types of systematic approaches when it comes to the above-mentioned problems [listed are urban renewal, environmental improvement, the overcoming of the nutritional problems of the world population, healthcare, and penal and legal enforcement, particularly in the context of application of computer technology; J. P.]' (Rittel 1992, 38).

12 'We are also suggesting that none of these tactics will answer the difficult questions attached to the sorts of wicked problems planners must deal with. We have neither a theory that can locate societal goodness, nor one that might dispel wickedness, nor one that might resolve the problems of equity that rising pluralism is provoking. We are inclined to think that these theoretic dilemmas may be the most wicked conditions that confront us' (Rittel and Webber 1973, 169). Despite having expanded planning theory to encompass second-order systems and despite the integration of intuitive phases into planning activities, Rittel had no genuine solution to dealing with 'wicked problems'. On the one hand, he claims with great definiteness that the principle of all planning can be characterised with a single sentence: 'Determine the components of the system to be designed, then link together their interdependencies and lay out the components in such a way that they do justice to the objective of the system.' On the other hand, we read just a few sentences later: 'Whence do problems arise in the first place? Who is authorised to formulate problems, and decides about the discussion and their inclusion in the agenda? This is readily answered: the process of problem formation is completely out of control, and to some extent even monopolised. I am thinking for example of the press, which has a disproportionately large influence. [...] Depending upon where the discrepancy lies, other problems arise, and there is no objective authority that is permitted to identify the real problems amongst these various possibilities. Everything depends upon how we explain things, how we see a problem, how we perceive the world progressing. These other so-called images we carry around in our minds; our image of the world: as it is, as it should be, as it will be, as it should not be, and so forth' (Rittel 1992, 69).

13 'Why are design problems indeterminate and, therefore, wicked? Neither Rittel nor any of those studying wicked problems has attempted to answer this question, so the wicked-problems approach has remained only a description of the social reality of designing rather than the beginnings of a well-grounded theory of design. However, the answer to the question lies in something rarely considered: the peculiar nature of the subject matter of design. Design problems are "indeterminate" and "wicked" because design has no special subject matter of its own apart from what a designer conceives it to be. The subject matter of design is potentially universal in scope, because design thinking may be applied to any area of human experience. But in the process of application, the designer must discover or invent a particular subject out of the problems and issues of specific circumstances. This sharply contrasts with the disciplines of science, which are concerned with understanding the principles, laws, rules, or structures that are necessarily embodied in existing subject matters' (Buchanan 1992, 16).

14 Not unlike Buchanan, I see that designers are able to explain their problems by situating the factors that condition the complex of a problematic situation, which is to say they capture and direct the accursed

or perplexing confusion in some way symbolically, diagrammatically, or in spatial models, achieving an overview and clarity through these operations. In contradistinction to Buchanan, however, I do not believe that such a localisation in relation to his schema of sign, object, action, and idea succeeds. I do not share Buchanan's idea of perceiving in the 'placement' of signs, objects, actions, and ideas a kind of style which allows designers to repeatedly solve problems and contributing to their characteristics. 'As an ordered or systematic approach to the invention of possibilities, the doctrine of placements provides a useful means of understanding what many designers describe as the intuitive or serendipitous quality of their work. Individual designers often possess a personal set of placements, developed and tested by experience. The inventiveness of the designer lies in a natural or cultivated and artful ability to return to those placements and apply them to a new situation, discovering aspects of the situation that affect the final design. What is regarded as the designer's style, then, is sometimes more than just a personal preference for certain types of visual forms, materials, or techniques; it is a characteristic way of seeing possibilities through conceptual placements. However, when a designer's conceptual placements become categories of thinking, the result can be mannered imitations of an earlier invention that are no longer relevant to the discovery of specific possibilities in a new situation. Ideas are then forced onto a situation rather than discovered in the particularities and novel possibilities of that situation. For the practising designer, placements are primary and categories are secondary' (Buchanan 1992, 13). More research is required in order to explain intuitive performance within the creative process that is touched upon with the 'doctrine of placements'. Required here is a more precise notion of how experience is registered in perception, perception into action, and action in turn into objects. Aesthetic experience strives towards holism, and is fed by the sources of everyday experience, sources that do not conform to Buchanan's schema (sign, object, action, idea) because it fails to detect their particular quality. Valuations proceed along highly complex, fraught processes whose characteristics consist precisely in the interconnection between action and object and its reality as a sign and its reality as a neuronal event. Additional allusions to this can be found further on in this text.

15 Nor do I regard the deepening of unclear information to be the *specific* object of design work. When designers are engaged in this regard (and often, they have no choice), they adopt the methods and tasks of other professions – i.e. discussions around strategic market development, research into the possible applications of specific technical principles, on the dimensional tolerance of materials in fabrication technologies, and so forth.

16 Unfortunately, the ability to dissolve existing orders is insufficiently appreciated. Often, creative personalities are regarded as significant only because they inject chaos and disorder into the system. Cf. Peter Kruse's notion of the function of creative people in delimitation from 'brokers' and 'owners'. Worth viewing in this context is the series of YouTube sounds Lutz Berger and Ulrike Reinhard have made with Peter Kruse in the sphere of SCOPE 08 (The Future of Learning + Working). Access may be possible at: http://www.youtube.com/watch?v=FLFyoT7SJFs&feature=player_embedded. Problematic however is Kruse's typology of broker, owner, creator, which is based on an analogy with the human brain (limbic system, cortex, and rising stimulation), mushrooms into a kind of doctrine of social typology.
On the other hand, my own proposal for positioning pattern formation at the centre (with Kruse, of 'change management' in the course of innovation processes, with me, of creative and 'aesthetic' cognitive achievement, to some extent parallel to Kruse's ideas, which return again and again to the relationship between stability and instability within great creative phases). Kruse focuses on the fact that new pattern formation self-evidently presupposes the destabilisation of old patterns. And it is here that he sees the designer or artist – or precisely the 'creative personality' at work: they provide disorder and chaos.
Crucial with Kruse's position, from my point of view (and with many other positions as well that have been addressed in the text above), is his tendency to recognise creativity as merely an intellectual phenomenon. Kruse also regards culture as a main thing where the habitual, something that is immaterial, is transmitted. In his conception, it seems to me, information is grasped, the something immaterial, not material, not embodied.
Only seldom, however, designers have been remunerated for what I have referred to as the 'dramatisation' of their work, the generation of vagueness, the deconstruction construction of order, although all of that is indispensable at the start of any work on form. Paid for as a rule is the establishment of form. Present in fact are deficits when it comes to reflection: designers take little account of their capacity to call into question the world as it is. Every new design dissolves precisely this process of deconstruction in a positive sense. Secondly, the psychological processes and cultural backgrounds which motivate their critiques are barely thematised. Because our understanding of design is positive in character,

everyone stares spellbound at the origination of reality, hardly at all at its critique. Critique is not solely rejection, although it does begin – according to Foucault's celebrated dictum – with a refusal to be ruled over (by circumstances). Invoking Kant, Foucault regards the concept of critique as being congruent with that of enlightenment (Foucault 1997, 41–82).

17 In fact, the qualities of the factors that determine form in a design project are so diverse and contradictory that their integration or synthesis is in any case always arbitrary. This arbitrariness belongs to the most conspicuous traits of design work, and makes it so desirable. It however also explains the provisional nature of all work on form. Nevertheless, the dissolution of existing form orders – as we learn from information theory – is also a formal act.

18 Here, I am moving entirely along the lines of the themes stated two years ago by Alain Findeli when, in contradistinction to Nigel Cross, he presented his interpretation of 'designerly ways of thinking'. Findeli too proposes that the application dimension of the object be integrated into the research. Findeli too believes that the specificity of design could well lie in its ability to show what has gone awry in the world. Findeli too believes that the designer's perspective of the world should not be descriptive, but instead diagnostic. For him, a fundamental task for the designer is to bring things that have gone awry in the world to a better state (Findeli 2008a). But the question remains: What are the criteria for 'better'? I attempt here to penetrate deeper into the design process itself in order to demonstrate how the vast critique of the 'awryness' of the world is also a critique of its aesthetic arbitrariness, and that it resides in the aesthetic dimension of design (Gestaltung) itself. Called for here is not some transcendent notion of harmony and beauty, which Findeli regards as necessary in some ideal way when he recommends that designers regard the world in the style of the ancient Stoics. According to Findeli, the awryness of the world reveals itself through reference to an ideal, systematic beauty and harmony. I do not believe in the success of holding up ideological images – no matter how beautiful they might be – to the world. Nor do I toss the materialistic, agnostic, and positivistic theories into the same pot of an objectivistic, positivistic scientific tradition.

I very much share Findeli's emphasis on the question of the specific object of design research (Findeli 2008b). Also his way of questioning the function of research ('Who is helped, what is researched in and through design?') is familiar to me. I regard the user, however, not simply as a 'recipient' or only as a passive receiver of design, as suggested by the 'Bremer model' (with a conceptual 'upstream' and a user-oriented 'downstream'), which is repeatedly overworked by Findeli. And for me, the 'anthropology of the project' (Findeli 2008a) remains extremely nebulous when it comes to defining what is specific to the design of an object. His claim that it is a question of the 'human experience to be in a design project' is a somewhat tautological response to the question of the specificity of design. Nor is this tautology mitigated by the deflection of saying that the 'specific human experience of being in a design project' is very similar to the experience of 'being in modernity' (Findeli 2008a). A text version of this lecture is available as a PDF at: https://www.researchgate.net/publication/235700599_Research_Through_Design_and_Transdisciplinarity_A_Tentative_Contribution_to_the_Methodology_of_Design_Research, reviewed last on 20 March 2019.

19 Designers achieve this disorganisation with the resource of objection. Their criterion is the insufficiency of the components found in a complex set of interrelationships and context. Coming into play here is the human individual and his aesthetic capacities for judgement, his readiness to endure continuous impositions. Expressed abstractly, the designers explode the obligatory quality of previously effective factors by calling the existing order and its degree of complexity into question. This can occur directly – then it is called hacking (see below). It can, however, occur indirectly, via the detour of diagnosis, research, critique, the present forms being translated by the critical actors (de-signers) into linguistic or image signs, diagrams, and spatial models. For such critical translation work and the configuration of a structure, designers like to use media in which structural relationships can be kept liquid and virtual. With paper, they prize the absence of the third dimension, and the possibility of overdrawing, and they flood their muting boards with Post-its or allow mind maps to proliferate. But design also increases or reduces the complexity of previous arrangements. In liquid and virtual media, they try out innovative localisations, vary hierarchies, associate elements from mutually distant contexts or those hitherto regarded as foreign, and transfer these clusters of factors into formal structures, bringing the factors in modules and their spatial orientation towards one another in materials, volumes, and colours. Because they reorder that which has been brought into a state of disorder, shaping it into a new whole, designers value the interrelationships between the elements aesthetically.

20 Horst Rittel was hardly the last or the only person to remark that some problems are so intricate and devilish that they evade all systematic calculation. According to Jeff Conklin's definition, 'wicked problems' are the ones we understand only once we have solved them. Conklin emphasises the incapacity of the problem-solver to explain the resolution process itself (Conklin 2005). For more information, see the CogNexus Institute website at http://cognexus.org. Conklin describes this design approach as 'opportunity-driven' (Conklin 2005, 5) and cites Raymonde Guindon, who calls it 'opportunistic' (Guindon 1990, 305–344).

When we connect these statements with the observations of creativity for research, we arrive at the following formulation: Whoever is in the flow cannot be interrupted in order to convey details to others. Here, the crucial point is precisely travelled from a state of non-awareness of a possible solution to ideational realisation and awareness of it in a single move. For more see Mihaly Csikszentmihaly's *Creativity. Flow and the Psychology of Discovery and Invention*.

That design deals with recognitional processes ranging from non-conscious to conscious, need not mean that this particular form of rationality is inaccessible. In this instance, however, enlightenment does not automatically mean methodological mastery over them. Wicked problems are different from tasks such as chess, Rubik's Cubes, or the inevitably rather dreary jigsaw puzzles, for which everything required for a solution is right in front of you. With wicked problems, it is not a question of making the correct decision within a well-regulated system in order to arrive at a successful conclusion. That is mere technique. Machines can do that as well. Which is ultimately why machines win at chess against humans. They are immune to the tactical and strategic pressure which human chess players deploy against one another. Machines are active outside the emotional calculations of human actors, which evaluate the pattern development of the antagonist empathetically and, through this experience of a specific situation, push the tempo and trigger errors.

21 This would mean understanding the aesthetic, the sensuous dimension of human experience, as a humane formulation of the technical, and not merely its decor or envelope.

22 Most recently in 'Nichts Neues. Keine Zeit. Einige Bemerkungen zu Gott und der Welt', in *Sushi* 12/2010, available for download at www.petruschat.com.

23 As early as 1959, the behaviourist Konrad Lorenz called attention to the fact that the structural organisation of data is capable of grasping and processing far greater quantities of information per time unit than linear, measured processes.

24 That embodied intelligence and its capacities for forming gestalts can respond better to complex challenges than linear automatism does not mean that gestalt formation does not encounter limits.

25 This is a process of transformations and transpositions. On the concept of transpositions, see my 2006 lecture 'Transsemantische Zustände', available for PDF download at www.petruschat.com.

26 'The complexity that we can manage unconsiously paralyzes us when we bring it to consciousness. If we begin to reflect-in-action, we may trigger an infinite regress of reflection on action, then on our reflection on action, and so on ad infinitum. The stance appropriate to reflection is incompatible with the stance appropriate to action. [...] That fear that reflection-in-action will trigger an infinite regress of reflection derives from an unexamined dichotomy of thought and action. If we separate thinking from doing, seeing thought only as a preparation for action and action only as an implementation of thought, then it is easy to believe that when we step into the separate domain of thought we will become lost in an infinite regress of thinking about thinking. But in actual reflection-in-action, as we have seen, doing and thinking are complementary. Doing extends thinking in the tests, moves, and probes of experimental action, and reflection feeds on doing and its results. Each feeds the other, and each sets boundaries for the other' (Schön 1983, 278 and 280). These remarks, however, refer to reflection of complexity within practical action itself. I am speaking here in particular of the unfolding of complexity that is evoked in the preliminary stages of 'actual' design work because it might be useful in solving problems. Nevertheless, an unfolding of complexity which precedes actual design work is thematised in the design process as well, and must be tamed by designers. As indicated above, I locate the resources for such a taming in the aesthetic dimension.

27 Whoever looks at an image perceives a selection of points on a surface. Whoever paints a picture brings out this wholeness successively through a linear implementation. But the motif that propels him is engendered by experiences of wholeness.

28 Out of the complex pool of data, the reality, which lies before our eyes, the organism only becomes aware of data which reaches it through the sensory apparatus. These physiologically anchored overviews also provide the selection procedures between ascending levels of mental processing with a basic framework.

29 It is here that intuition comes into its own. I do not want to celebrate the irrational. And I certainly believe that intuitive processes are accessible to rationality. Here, I simply want to mention that the neurophysiologist Antonio Damasio was able to demonstrate how intuitive processes are able to resolve successfully highly complex situations through decision and action by employing somatic markings, which is to say embodied experiences, in order to order the factors of the problem space. For more see Demasio's *Descartes' Error: Emotion, Reason and the Human Brain*, in particular chapter 9 on the 'Somatic-Marker Hypothesis'.

30 I agree with much of what Pieter Jan Stappers has to say about the integration of research in the design process; in one respect, namely the one I discuss here, however, I do not share his point of view. I see the effectiveness of the evaluative functions not, as he suggests, simply on the side of rational science and its testing methods, but also and precisely on the side of designers and their emotional intelligence. Precisely this ability to arrive at the significance of factors via sensory distinctions is the special feature of aesthetic valuation, upon which the specificity of the designer's capacities is based in contradistinction to those of the engineer. This valuation, meanwhile, is now a process that can be articulated verbally; instead, this sensuous valuation of factors takes place through the process of generating form *(Formgebung)*, in the emergence of the gestalt *(Gestaltwerdung)*, in decisions about which details are emphasised, reworked, contrasted, related to one another, and so forth. The evaluation of factors can take place in different forms, pointing to economic, entropic, climatic, and other criteria, but cannot result in a gestalt when the aesthetic dimension is excluded. Secondly, Stapper's model of the design process: for him too, the process culminates in a product. This means that the intersubjective character eludes him, and he therefore does not really succeed in allowing for criteria of design valuation which have become independent of the merely rational modes of evaluation found in the 'sciences'. Then all that remains is the mere statistical tests, and hence the redundancy in relation to what exists and is expected. My position on this is explained above in the text. For more, see Pieter Jan Stappers's chapter 'Doing Design as a Part of Doing Research', in *Design Research Now: Essays and Selected Projects* (2007, 81 ff).

31 See John Dewey's concept of aesthetic experience in *Art as Experience* (1958, 33 ff).

32 See my remarks on Walter Zeischegg in 'Befreit die Technik und Ihr befreit die Form', a text on Max Bense and Walter Zeischegg, in the 'Ulm issue' of *form+zweck* 20/2003, available for download as a PDF at www.petruschat.com.

33 On the problem of complexity and complexity reduction in the context of design, on the relationship between aesthetic abilities and formation of consciousness, see my article 'Das Leben ist bunt. Einige Bemerkungen zum Entwerfen', in *form+zweck* 21/2005, available as a PDF download at www.petruschat. com. I rely here upon the research of Edelman and Tononi's *A Universe of Consciousness: How Matter Becomes Imagination* (2000).

34 By culture I mean the (objective and subjective) conditions that are given for the individual, and from which something special can be created. With this recourse to individuality, I anchor the concept of culture more deeply in terms of developmental history than the concept of technology, whose differentiation against nature (its development) is a cultural achievement.

35 Many semiotic explanations boiled down to saying that designers must deploy signs in order to ensure redundancy in their designs. This compels cultural studies – after the model of hermeneutics – to determine which configurations are accessible to which cultural interpretations.

36 That which is chalked up to implicit knowledge rests for the most part upon strategies for mastering complexity which have become submerged in preconscious levels and – sometimes unexpectedly – resurface in very different (in unexpected problematic) constellations which challenge subjectivity.

37 At this point, there is not enough time to elucidate the difference between the proposal for seeing the point of departure for new problems in the product, and for example the proposal also offered by Alain Findeli, to incorporate the aspect of use into research into design objects. Findeli's proposal is a genuinely good pointer for extending research in the direction I would like to see it moving. For me, however, his approach falls short because he calls 'the right side', which aims towards use, simply 'reception'. This only extends the metaphor of the user as a kind of shopping basket or bag with special claims. This still resonates with the old theme of spelling out the design process as a sender-receiver constellation. This pushes the user into a passive role from the outset, making it difficult to address the question of how users can be integrated more actively into project development. The political dimension of design is thereby reduced to education or delectation *(Beglückung)*. And the new life requirements of the user to participate in and contribute to decisions about resource utilisation, to become sovereign and 'creative', are hence not thematised at all, or insufficiently.

38 Production and consumption can be related to each other however only cyclically, that is to say in terms of a theory of reproduction.

39 See Richard Buchanan's proposal in 'Strategies of Design Research: Productive Science and Rhetorical Inquiry', in *Design Research Now: Essays and Selected Projects* (2007, 55 ff). An earlier attempt in the German cultural context *(Denkraum)* to take up rhetoric for an investigation of the design process can be found in Isabella Sladek's 'Die Bildsprache der Werbung', in *form+zweck* 5/1986; *form+zweck* 6/1986 and *form+zweck* 2/1987. Also profitable is an analysis of Vitruvius's *The Ten Books on Architecture*, where Caesar's admiral and architect master was apparently only able to investigate architecture, whose functions were unknown to him, and which lay ruined around him, with the resources of rhetoric, and in the form of readings.

40 At the human level, *Gestaltung* rests on cultural requirements. *Gestaltung* is the formation of identity, mediated through the forms of objects. In a certain sense, the stability of culture applies to organisms as well. For machines which also possess design potential, at the current technological level, I would refrain from associating the requirements of the self-model with the concept of culture. And this form of identity-formation, mediated through objects, is also based on cultural requirements.

41 It is certainly important here to recognise and to consider the semiotic dimension of design processes. The difficulty here consists in recognising the way in which actors become conscious of neuronal activity only in the signifying act (to become a sign) and not – as is usual in theories of the sign – to merely comprehend signs as pre-existing elements, to describe their combinatorics, to generate analogies between *Gestaltung* and rhetoric, and hence merely to capture redundancy in relation to the existing culture.

42 To make certain that the readers of the German-language *Die Welt* understand him correctly, Mateo Kries appends the following lines to the title of his remarks: 'The authority of the designer threatens to become a dictatorship: alongside clothing, cars, and furniture, they have long since begun to design the life of tomorrow' (Kries 2010).

References

Buchanan, R. (2007). 'Strategies of Design Research: Productive Science and Rhetorical Inquiry'. In Michel, R. (ed.). *Design Research Now: Essays and Selected Projects,* Board of International Research in Design (BIRD). Basel: Birkhäuser.

Buchanan, R. (1992). 'Wicked Problems in Design Thinking'. *Design Issues*, vol. 8, no. 2 (Spring, 1992), 5–21, also available at http://www.jstor.org/stable/151163.

Conklin, J. (2005), 'Wicked Problems and Social Complexity'. In *Dialogue Mapping: Building Shared Understanding of Wicked Problems*. Chichester: John Wiley & Sons.

Cross, N. (2007). 'From a Design Science to a Design Discipline: Understanding Designerly Ways of Knowing and Thinking'. In Michel, R. (ed.). *Design Research Now: Essays and Selected Projects,* Board of International Research in Design (BIRD). Basel: Birkhäuser.

Csikszentmihaly, M. (1996). *Creativity: Flow and the Psychology of Discovery and Invention*. New York: HarperCollins Publishers.

Damasio, A. R. (1994). *Descartes' Error: Emotion, Reason and the Human Brain*. New York: G. P. Putnam's.

Dewey, J. (1934/1958). *Art as Experience*. New York: Capricorn Books, G. P. Putnam's Sons.

Edelman, G. M., Tononi, G. (2000). *A Universe of Consciousness: How Matter Becomes Imagination*. New York: Basic Books.

Eisenstein, S. M. (2006). *Jenseits der Einstellung. Schriften zur Filmtheorie.* Lenz, F., Dieserichs, H. H. (eds.). Frankfurt am Main: Suhrkamp.

Findeli, A. (2008). Keynote at Q&H Conference '08, Searching for Design Research Questions, audio file accessible at http://www.designresearchnetwork.org/drn/content/q-%2526amp%3B-h-conference-%2526%2523039%3B08-keynote-alain-findeli-searching-design-research-questions last reviewed on 25 January 2011.

Findeli, A. (2008). 'Research through Design and Transdisciplinarity: A Tentative Contribution to the Methodology of Design Research', at the conference FOCUSED – Current Design Projects and Methods, held in Switzerland in 2008.

Foucault, M. (1997). 'What is Critique?' (original: Qu'est-ce que la critique?). In *The Politics of Truth*. Los Angeles: Semiotext(e), 41–82.

Guindon, R. (1990). 'Designing the Design Process: Exploiting Opportunistic Thoughts'. *Human-Computer Interaction*, vol. 5, 305–344.

Hayes, J. R., Simon, H. A. (1974). 'Understanding Written Problem Instructions'. In Gregg, L. (ed.). *Knowledge and Cognition.* Potomac, MD.: Lawrence Erlbaum; cited from the reprint in Simon, H. A. (1979). *Models of Mind*. New Haven: Yale University Press.

Kries, M. (2010). 'Wollt ihr das totale Design? Die Herrschaft der Gestalter droht zur Diktatur zu werden: Neben Kleidern, Autos und Möbeln entwerfen sie längst unser Leben von morgen. Plädoyer für eine neue kritische Designtheorie'. In *Die Welt*, 20 April 2010.

Kruse, P., Berger, L., Reinhard, U. SCOPE 08 (The Future of Learning + Working), http://www.youtube.com/watch?v=FLFyoT7SJFs&feature=player_embedded.

Newell, A. (1972). *Human Problem Solving.* Englewood Cliffs: Prentice-Hall.

Petruschat, J. (2011). 'Fassungslosigkeit. Einige Bemerkungen zum freien Spiel der Kräfte', lecture, Zurich.

Petruschat, J. (2010), 'Nichts Neues. Keine Zeit. Einige Bemerkungen zu Gott und der Welt'. In *Sushi* 12, PDF download at www.petruschat.com.

Petruschat, J. (2006). 'Transsemantische Zustände', lecture, Stuttgart, PDF download at www.petruschat.com.

Petruschat, J. (2005). 'Das Leben ist bunt. Einige Bemerkungen zum Entwerfen'. In *form+zweck* 21/2005, available as a PDF download at www.petruschat.com.

Petruschat, J. (2003). 'Befreit die Technik und Ihr befreit die Form'. *form+zweck,* 20/2003, available for download as a PDF at www.petruschat.com.

Popper, K. R. (1999). *All Life is Problem Solving.* Oxford: Routledge.

Rittel, H. W. J. (1992). 'Zur Planungskrise: Systemanalyse der "ersten und zweiten Generation".' In Rittel, H. W. J. *Planen Entwerfen Design. Ausgewählte Schriften zu Theorie und Methodik,* Berlin: FMI Facility Management Institut Forschungsgesellschaft mbH &. Stuttgart, Berlin, Cologne & Kohlhammer: Wolf D. Reuter, 37–58.

Rittel, H. W. J. (1967). 'Systematik des Planens'. In Rittel, H. W. J. *Planen Entwerfen Design. Ausgewählte Schriften zu Theorie und Methodik.* Berlin: FMI Facility Management Institut Forschungsgesellschaft mbH &. Stuttgart, Berlin, Cologne & Kohlhammer: Wolf D. Reuter, 63–73.

Rittel, H. W. J., / Webber, M. M. (1973). 'Dilemmas in a General Theory of Planning'. In *Policy Sciences,* vol. 4, no. 2, 155–169.

Schäffner, W. (2010), 'The Design Turn. Eine wissenschaftliche Revolution im Geiste der Gestaltung', in Joost, G., Mareis, C., Kipple, K., *Entwerfen Wissen Produzieren. Designforschung im Anwendungskontext.* Bielefeld: Transcript.

Schön, D. A. (1983), *The Reflective Practitioner: How Professionals Think in Action*, New York: Basic Books.

Simon, H. A. (1996), *The Sciences of the Artificial*, 3rd edition, Cambridge, MA: MIT Press.

Simon, H. A. (1996) *The Sciences of the Artificial*, 3rd edition 1996, Cambridge, MA: MIT Press, here Chap 5, 'The Science of Design, Creating the Artificial', pp. 111–138.

Sladek, I. (1986–87), 'Die Bildsprache der Werbung', in *form+zweck* 5/1986; *form+zweck* 6/1986 and *form+zweck* 2/1987.

Spillers, William R. (ed.) (1974), *Basic Questions of Design Theory,* Amsterdam: North Holland Publishing Company.

Stappers, P. J. (2007), 'Doing Design as a Part of Doing Research', in Ralf Michel (ed.), *Design Research Now: Essays and Selected Projects*, Board of International Research in Design (BIRD) Basel, Boston & Berlin: Birkhäuser Verlag.

Winnicott, D. W. (2005), *Playing and Reality* (1971), London and New York: Routledge.

AN UNFOLDING POLITICAL AGENDA

Tony Fry

No matter who we are, we all exist in an age of post-political party politics. For functional nations, formal politics has become the preoccupation with the management of the national economy, national security, and national interests (inflected by varying degrees of nationalism). In contrast, for dysfunctional nations ('failing or failed states') it seems that expediency and chaos now mostly determine the nature of everyday life. The political machine is broken, or corrupt, or both.

It is against this schematic contextual background that an argument for design as politics will be made. This argument will make the difference clear between it, as an advocated position, and the use of design in supporting political ideologies, parties, and movements.

Over the duration of its earthly habitation our species, Homo sapiens, created a world of human construction within the biophysical world of its dependence – now named as the anthropocene. This constructed world arrived by the exercise of human agency by design as an anthropologically inherent feature of our being's propensity to prefigure its actions, and by it later becoming consciously applied practice. Designing existed before design. By implication, at any given time, whatever was brought into being had a direct consequence on the form of the future. The efficacy of all things designed is that they go on designing, with positive (futuring) or negative (defuturing) consequences. At their most basic, the futural qualities of the things that constitute the designed environment are evident in their material impacts upon the natural world, seen in the degree of structurally inscribed unsustainability within the life of, and in, the anthropocene. One can thus conclude that the agency and consequences of designing and the designed are intrinsically political as environments and lives are changed. But in so saying a distinction needs to be more clearly made between politics and the political.

What has been brought into being by design, the impacts and futural consequences (the ontological effects), all result from the combined agency of the already designed and materialised, design practice, the establishment of relations between 'things', and the exercise of multiple structures of power – this as they all converge to support and express a claiming of the action of designing and the design being political. So framed the political agency of design creates an operational regime of order, or disorder, of an assembly of material things, by literally establishing an ontologically designing 'environment of our being in place'. Here then is what can be taken to be the long-standing ground out of which politics arrived. In particular, the very establishment of human settlement marked a fundamental transformation of human ontology. The Greek *polis* provides a good example of such a relational foundation wherein the material form and social order of the city were indivisibly connected and political. With the rise of modernity and the creation of colonial cities, exactly the same ontological ordering of space and everyday social life was made an operationally directive of urban form, but with very different consequences.

As we are seeing, the coming of the city, due to its social and economic complexity, gave the ontology of the human species a more overt political form. As the scale of this complexity of constructed worlds and social orders increased so did the political ontology of our being. The rise of the nation state, industrial society, and a geopolitical order all manifest and mark this development, all of which added additional layers to the onto-political condition of everyday life, this for the modernised and their colonised others. Moreover, in this setting the rift between politics and the political started to form. In summary:

Politics is an institutionalised practice exercised by individuals, organisations and states, while the political exists as a wider sphere of activity embedded in the directive structures of a society and in the conduct of humans as 'political animals.' Politics effectively takes place in the sphere of the political wherein the agency of things – material and immaterial – is determined and exercised as they are perceived, and become directly or indirectly influenced, by a political ideology. There has been a general societal perceptual failure to distinguish between the political and politics, in large part because, as Claude Lefort has pointed out, the latter acts to conceal the nature of the former (Fry 2010, 6).

As for currently existing democracy its politics cannot be accepted as the end point of political philosophy. Democracy is not, and so far has never been, fully realised. The voice of 'the common good' remains silent and unrepresented. More than this, democracy in the developed 'democratic world' has degenerated into televisualised 'consumer democracy' as directly connects to the notion of 'consumer sovereignty' (the politics of meeting consumer demands). As C.B. Macpherson argued several decades ago, this turns democracy into a commodity and this is one of the distinctive features of capitalist democracy (Macpherson 1973, 79). What it appeals to are base interests (like economic self-interest and nationalism), consequently voting is abstracted from what actually requires to be decided (especially in and for the long term). It does not rest upon or enable informed choices. Moreover, the exercise of decision is frequently undertaken via executive power and enacted by the bureaucratic regime of government. So rather than offering fundamentally different political options, visions, and directions, what is put on offer for capitalist nations are varied management options centred on upholding the economic and social order that underpins hegemonic capitalism. Here then is a politics in which political difference is neutralised and governments become administrators of public, and increasingly private, life. As such, governments act as the defining agent of which life has, or does not have, a value.

Subtending the illusory representative political process of democracy is also an authoritarianism exercised in the name of national security, national interests or state law – these domains are effectively placed outside the sphere of public influence. Moreover in 'an age of the threat of terror' everyone is viewed as a suspect, which means surveillance becomes visible and invisible presence in the public and electronic domain (with such action claimed to be the price of freedom). The creation of both fear and paranoia, have become endemic to mainstream politics as a means to extend restricted freedoms without popular resistance. The distinction between 'liberal' democracy and repressive regimes is often not what it seems to be. In the name of protecting it, freedom is taken away and mechanisms of control arrive. The distinction between the protection and management of society blurs.

Against such a background one can say that tectonically, the very ground of politics and the political is shifting. Not least as the technological constitution, transformation, and ontological designing of the anthropocene becomes ever more complex. Effectively, the rift between politics and the political becomes increasingly wider. At the same time the combination of the rapid growth in the human populations and the arrival of technological society as hegemonic, acts to make the worlds of human dependence far more unsustainable. A point has now been reached wherein the complexity of these worlds is now disjunctural from national and geopolitical orders of institutional politics. What this means is that formal political agendas, issues and forms of political leadership that preoccupy politicians, are structurally unable to recognise the relational complexity of the material and immaterial 'political nature' of the antropocene – including the synthesis of the natural and the artificial, together with changes taking place in the very valence of the emergent plural ontological forms of (post)human beings. In fact in so many ways, the political contestation of the future of humanity is being passed over by institutional politics. It is simply does recognise, and so is not attuned to, the level of fundamental worldly change under way. The same observation can equally be made of the design professions and designers of all stripes bonded as they are to servicing the status quo.

While design is not usually linked with the political, other than as a service provider for political parties, organisations and governments, and policies, it is, as is being argued, profoundly political in determining the futural consequences of whatever is designed as it goes on designing.

As the history of architecture and design confirms: cities, hospitals, prisons, offices, factories, homes, parks, public transport, utilities, infrastructure, public information and so on, all arrive with forms lodged in particular sets of ideological values that are predicated on how human beings should be viewed and treated (Fry 2010, 6).

Here again design (mostly in anonymity) is being essentially seen politically as it expresses and serves the power that directs, materially and immaterially, those ontologically forms that prefigure everyday life and the *made* world around us.

In the disjuncture between institutional politics, worldly complexity, and unsustainability (which includes extant and emerging modes of human high-consumption ways of life), there is a clear imperative for design to become fully recognised as an overt and potentially astute political practice. What this means is establishing a praxis that makes forms of the unsustainable present in order to make them foundational problems available for rigorous inquiry. The presumption here is not that design immediately solves the problems so exposed. We partly live among the insoluble: which means learning how to adapt and live with such problems. Equally there are problems that can be solved, but what this implies is the delivery of solutions at an absolutely fundamental level. Such problem solving obviously has to be beyond (i) the trite rhetoric of 'design as problem solving', which is so often a superficial meeting of a client's need and is often (and mostly unknowingly) generative of problem creation; and (ii) 'wicked problems' that can have an embedded problematic of problem definition. These remarks beg a few qualifications.

First, what is being said is suggestive of a new design practice where the need to understand what can and should be designed overrides acting on instructions (from a brief/client). Next, there is no presumption that the identification of a problem (even when engaged) is going to be met with a design solution. Finally, design so viewed begs to be seen as a mode of ontological inquiry of the ontology of the unsustainable (which implies that unsustainability is not self-evident, fully understood, or being sufficiently interrogated).

Certainly unsustainability cannot be reduced to merely biophysical condition characterised by the likes of the impacts of climate change and its associated environmental crises (including the loss of biodiversity folding into the indication that the sixth mass extinction event has commenced, natural resource depletion and stress, populations pressure, pandemics, and food security). Other issues include population displacement and related conflict; the potential for global nuclear conflict; and the psycho-ontological and worldly impacts of the technological transformation of 'the human' (already producing species fragmentation, creating dangers of the totally instrumentalised subject, while 'affording' an enhanced ability for the privileged to be technologically sustained). New forms of inequity emerge and a possible major collapse of the global economy (with massive broad impacts beyond the economic). This is not to say all these 'disasters' will arrive, but it is to say that it is certain that some will.

Risk identification is indivisible from avoidance action. Such action again folds back into a new design agenda wherein design praxis is constituted and mobilised to accept a greater responsibility for what design has brought into being. Bluntly this will not happen unless designers are far more adequately educated – this not to better design but to be more capable of grasping: (i) the nature of the worlds and conditions of unsustainability in which the designed now arrives and, (ii) what the designed designs in a worldly sense, and thereafter what needs to be designed, or not. Obviously design as political now transmutes into the politics of a transformation of design education. Unavoidably the 'design community' (all of whom have a vested interest in some way in design) at some point need to make a choice: economic self-interest and predominantly its links to an economic service provision, and so to the unsustainable status quo, versus making a major move into a nascent sphere of design as a counter-political practice with an ability to challenge forms of excessive economic self-sustainment. Can this be done? The answer, as will be argued, is yes but in action not voice, and not in one move or instantly.

A Question of Transition

Design practice has to be grasped, presented, and taken well beyond the rubric of the order of activities of current design professions. Designing goes ahead of all that we intentionally bring into being, thus it is integral to all forms of prefigured human productive action. This means it has to be liberated from design as 'the design of a design'/a 'designated design object' – this so it may thereafter be seen in terms of the consequence of much that comes into being by design (material and immaterial as designed things at large, and as they functionally/operatively, go on designing, with positive or negative consequences for people, societies, environments and the biosphere in general. Notwithstanding such remarks, it has proved enormously difficult to gain widespread recognition of the importance of design within and beyond the design professions. This is a situation made worse by the restrictive practices and perceptions of the design professions themselves as they strive to claim ownership of design and be its authoritative voice. The way the media dominantly presents design, reducing it to aesthetic concerns, with a focus on taste and style, adds to the problem – a situation that so often design practitioners are complicit with.

Such seemingly apolitical actions actually have significant political consequences as they act to trivialise general perceptions of design and reinforce the delimited view of it. Thus, by implication, they negate attempts to establish a more substantial, social, broader, and more rigorous understanding of design, designing and the designed – one that recognises just how much design is a determinate force upon the fate of the world upon which we humans, and other species, depend.

Clearly the transitional challenge to be embraced goes to the remaking and extension of design practice itself, design education, and thereafter to how design is to be futurally understood in contemporary societies beyond its currently limited aesthetic, instrumental, and economic characterisation. Obviously this challenge carries an enormous research agenda within it. The design professions identified that the current unambiguous bondage as a service industry to the economic status quo needs to be broken. This is not to suggest an unrealistic abandonment of the economic, but rather a more substantial, critical, and transformative relation to it.

At the same time an independent activity distant from the extant economic sphere needs to be established. These remarks again echo the case for the politicisation of design and the prospect of a much extended design research agenda.

Placed in its largest frame, what is being argued for is the redirection of design against all dimensions of the structural unsustainable and towards the advancement of Sustainment (a post-sustainability intellectual project and transformative process).[1] To do this is to position design with ability to be ...

[...] truly futural and political – which implies design becoming more dynamic, more powerful and more able to communicate the significance of designers to society in general. This means that the way designers think, the culture they create and the practices they establish have to radically break with existing and dominant patterns (Fry 2010, 76).

The development of a viable political theory of Sustainment is one thing, the design and creation of the institutions to materialise it is another. Clearly between this theory, and its institutional realisation, a means of material enablement has to be created. The project of Sustainment carries the beginnings of such praxis.

'Redirective practice' is a meta-practice that can be understood as a gathering of practices under a common understanding centred on a relational theory of knowledge (counter to disciplinary divisions of knowledge and practice) within which design is situated as a powerful praxeological actor. In its emergence, 'redirective practice' acknowledges the imperative of dealing with 'what already exists' in order to turn it towards the future with a sustaining ability. Positioned between things as they are (the structurally unsustainable as normalised) and Sustainment, 'redirective practice' is about the creation of futuring potential. Forms of redirection are myriad, including first of all a post-disciplinary redirection of the specific design practices, followed by actions such as the recovery and remaking of the discarded and the broken, as well as the repurposing of structures and products.

In the company of 'redirective practice' there is another crucial and still inchoate design practice: 'elimination design' (Fry 2008, 72–80). Its task is of equal importance to the dominant rationale of design: the bringing of new things into being. In this context 'elimination design' is an activity directed at finding and using ways to destroy the unsustainable and to respond to the need to create strategic means of constraint in an age of globalising and uncontrolled hyper-consumption (Fry 2010, 244–245). Connected to elimination design is 'no design-design' where the design act itself is eliminated (this implies a design solution in which design does not bring anything into being). Before going further, a little needs to be said about things (as designed object).

Things are not divided from us, or each other, by either organising principles, inertia, or social and material forms. Rather they operatively flow into each other, are configured in dialogue and are the resultant matter of discourse. To reiterate: 'we' bring designed things into being and they contribute to our becoming. Our habits, appearance, skills, health, ability to travel, play, ability to be entertained, eat and so on cannot be divided from things as they exist functionally, aesthetically, or ontologically. It follows, as now widely recognised by for example 'assemblage theory', 'actor-network theory' and 'object-oriented philosophy', that things have enormous agency (seen here in their politically designing significance) as they dispose us towards the world and so determine our impact upon it (be it to a lesser or greater degree as futurally negative or positive). The extent to which we are individually or collectively unsustainable cannot be divided from what we create and then how we use it and in turn how it uses us. The political as intrinsic to the nature and propensity of things may seem far removed from an association with 'the everyday politics of now'. Yet political regimes manage the economies of consumer society in ways to contribute to the generation of national wealth (not least by taxes), but without much knowledge or due regard for the designing consequences of consumption. Certainly they act with no ontological comprehension of the complexity of the ontological designing qualities of structure, products, and services.

What 'redirective practice' and 'elimination design' share is a recognition that there is a massive and much neglected task of dealing with what already exists as it fosters and extends the unsustainable. Much needs to be destroyed by design. For instance, consider the huge amount of what fills the shelves of supermarkets has little nutritional worth, is even harmful to health, or has no use value – thus it is just landfill waiting to happen (its fate in a matter of days or weeks), as is much else the retail sector puts into the marketplace.

Although again, the conceptual challenges are daunting and practical difficulties formidable, a capability is already starting to arrive to bring elimination design action into wider use. Yet it begs to become a major design development and done with far more rigour (and so can be viewed as another significant research need). Of course such action travels in two directions: towards the already existing unsustainable; and towards preventing the unsustainable coming into being. Linking both practices is a disposition towards designing that orders three practitioner priorities: design that eliminates the designing of anything material (to conserve especially non-renewable resources); redesigning the already designed (to eliminate the need for the new); and designing whatever new is to be brought into being in time (to slow down the production cycle and contribute to building a 'quality over quantity economy'). Framing all these activities is a requirement for a very clear understanding of the unsustainable at a material level as well as in terms of human values and conduct, plus psychologies (not least in relation to desires). There is obviously a political imperative to be registered: one to direct a far more serious engagement with the unsustainable which, at the most general, can be linked to 'a remaking of the institutional and political domain of design'.

Out of this activity of remaking comes the direction of the foundation of all design practices that are predicated upon the advancement of Sustainment. Unless this happens, humanity in all its current and mutating forms does not have a viable future. Clearly this is an intergenerational action determining that any really serious act of critical reflection will be concluded to be unavoidable. This action is now either being denied, or tokenistically addressed (as the instrumentalism of 'sustainable design' exemplifies). Indivisible from design practices remade is, as argued, the redirection of design education so that designers are sufficiently educated to understand the futural importance of design, the related responsibility that travels with designing and, above all, the nature of the complex world in which the designed arrives at and goes on designing in. This itself means a fundamental redirection/re-education of the vast majority of design educators.

Unambiguously such proposals are profoundly political, as has in part been elaborated elsewhere (Fry 2010). They are being introduced in modest forms in a wide variety of institutions. The more they are adopted, the more they will be met with resistance to overcome.

To ground much of what has been said, comment will be made on two projects that link to my own practice – 'The Studio at the Edge of the World'. The studio is a post-disciplinary design education and research entity that presents an implicit critique of design education, and tertiary education in general, while developing and researching projects that address situated examples of engaging 'the structural unsustainability of the contemporary world'. While the claim of these projects is modest, they do mark more than starting points.

Project One: Designing-in-Time and the Pearl River Delta

This project illustrated an example of 'designing-in-time' (a means of engagement with a problem that is yet to arrive). It was conducted with most mainland Chinese students in a Hong Kong Polytechnic University in 2005. Its situated focus was on the Pearl River Delta (PRD) of Southern China and the projected impact of sea level rises on the delta due to climate change – the region is recognised to be one of the most at-risk locations in the world. It is low lying, fed by three major river systems with a vast network of tributaries. At the same time, it enfolds the major cities of Guangzhou, Shenzhen and, at its mouth of the delta, Hong Kong. The PRD has a population of over 60 million people and is the most urbanised area on the planet (according to the World Bank). Economically it is not only an important manufacturing region, but is a key element of the largest economic mega-region in China. So what happens to the Pearl River Delta will not only have a massive impact on China, but also on the world in general.

The essence of the project was to employ a 'design-in-time' method based on designing back from the future scenarios. Central to this method, which my practice has successfully employed in award-winning urban design projects, was a comprehensive research exercise (within the time constraints of the project) that gathered geographic/environmental/climatic, economic, educational, historical, infrastructural, and demographic data. This material was analysed and connections were made (within the limits of the time frame of the project). Nonetheless, gaining this worldly knowledge provided a basis for multi-layered scenario building within the method (registering environmental event, social and economic impacts), and design responses (both material and immaterial) – all this was written into a graphic, annotated and linear narrative working back from the future (from between fifty and one hundred years) to the present. From plotting projected events and impacts design responses were added. Such an approach is clearly not about acting with certainty, or delivering resolved design solutions. Rather it is about three things:

- learning how to construct a relationally connected picture of design determinates (in contrast to 'designing in context');
- designing in the medium of time (in contrast to designing in space as exemplified by 'master planning'); and
- designing towards strategic decision making based on speculative impact evaluation across all scenario levels (in contrast to designing for 'market demand or user needs').

This method (only outlined here) to be fully realised is beyond the scope of a single and brief project. It actually registers a substantial programme of work that again directly confronts the nature of design education and practice. Rather than privileging 'how to' design it goes to the imperatives of 'why design'. It also returns to questions of 'what now needs/does not need' designing. In doing so it places worldly understanding, research, and critical evaluation at the front end, and in the core, of design education and practice. In turn it does this within the identified design ethos of 'redirective practice', 'elimination design' and 'no design-design' being dominant over the design of the new. Such change in education and practice would of course have the consequence of transforming what was being newly brought in to being design. It would equally effectively slow down the process of incremental change of manufactured products dramatically, extend the design life of what was produced, and help redirect a defuturing (global) economy towards the process of Sustainment (thus buying time in the unavoidable face of entropy).

Project Two: Responding to Unsettlement

This project was (and its afterlife is) about the concept of unsettlement as a basis of new project generation. It is again also linked to design education. To make sense of it we need to first outline how unsettlement is understood.

At its most basic, unsettlement is a disturbance/disruption in the current dominant modes of humanity's being-in-the-world as experienced and enacted individually and collectively.

Unsettlement, in sum, combines naming a moment in (and of) human existence that recognises an ending without any clear sense of beginning accompanied by a certain feeling of foreboding – a foreboding about the arrival of 'natural', or 'human induced', unwelcomed defuturing events in the immediate or distant future no matter whomsoever 'you' are or wherever 'you' are situated. Confronting such prospects shifts unsettlement from being thought of just as a 'state of the world' to it being lived as condition directed by a 'state of mind' (a psychology).[2]

Unsettlement fundamentally shifts the ground of desire and design. It does this by exposing how tenuous our 'grip on the world' actually is and in this context how pathetically misplaced most object-thing-based design 'solutions' are.

The causal conditions of unsettlement, as we shall see, can come from a convergence of many forces and associated threats. What unsettlement predominantly does is to amplify the biological, geopolitical, and psycho-social instability to which the planet's population is, and increasingly will be, exposed. As such it marks an emergent and fundamental change in the level of nihilism lived and felt by vast numbers of the global population. In this situation so many values become devalued, and all sense of agency is reduced or lost – an expectation of a coming maelstrom also, and slowly, forms.

The project that centred on unsettlement was structured around a five-week studio 'learning by practice' event.

The event focused on gaining a comprehensive and theoretically informed understanding of structurally inscribed unsettlement, recognising that unsustainability in fact breaks the world – the world of our fundamental dependence. But the world of biological life has been broken before. In fact there have been five major planetary extinction events. Each time the biomaterial world remade life out of the post-destruction remnants. In the most serious of these events, some 253 million years ago, over 90 per cent of all living matter was destroyed. 'We' are a product of the life matter that was left over. It is with some irony then that there is now talk, within the scientific community (based on the rapid rate and scale of the loss of biodiversity) that the start of the sixth extinction event has now commenced: this as a result of the sum of our own defuturing actions. We are the breakers. Thus unsustainability rests with us, and knowing this is profoundly unsettling.

The 'world' as evoked by our presence is an environmental, conceptual and material construction of our own creation existing in relation to the given biophysical world of our dependence. Both have become essential conditions that need to be sustained for us to continue to be. Our created world – the anthropocene – the world-within-the world – continually designs the 'nature' of our human being.

Unsustainability comes from the unchecked consequences of anthropocentrism enacted via the formation and agency of the anthropocene. We, as 'the most dangerous of animals', are without constraint. The more of us there are, and the more technology has accelerated the speed and volume of appropriated world resources, the more our destructive power has increased. Somehow for all our 'cleverness' an essential stupidity endures.

In sum, notwithstanding all the manifestations of a broken world (environmental and climatic, proliferating conflict, hyper-consumption, mountains of waste, excesses of wealth and poverty), they are not causes of the unsustainable. Rather they are its symptoms. The actual fundamental cause is us and our mode of earthly habitation.

Against this backdrop the Learning Event looked at examples of the visualisation of intercultural action in the age of unsettlement. This was done through group projects that provided the means to engage unsettlement in situated contexts (storytelling; a commercial constructed social space of encounter in the city; and a confrontation with an environment of conflict). In so doing it acknowledged the unfolding of the relational complexity of nature of concepts recognising that *'we'* can't *'solve a problem with the thinking that created it'* (Einstein), moreover we live in a world where many problems exist for which there are no solutions (although we can only conclude this with the limits of our current level of knowledge – the certain may be uncertain). Existing thinking, as has always been the case, is going to be disrupted and overwhelmed. Likewise, necessity may drive human beings towards other ways of thinking that go beyond the limits of the hegemonic instrumentalism of the present, but again this is not certain.

A transformation of thought cannot become an immediate reality. It only arrives as 'an opening' from a determined intellectual investment and community of thought over time. In this context, language was used in the Learning Event to unsettle unsettlement (as a defamiliarisation of the familiar). Clearly this created a need for a communication practice that understood just how hard it is to communicate – not least with that which is new, unfamiliar, and challenging.

From working in groups, participants then moved on to develop individual projects.

Thereafter, a modest exhibition was designed and produced, to which an invited audience came. It presented both the individual and collective projects. The underpinning rationale of the entire learning exercise was in fact to place the participants in a position to author their own projects dealing with unsettlement (including their own existential unsettled condition). Effectively this meant that the most important aspect of the event became what happened after it.

The key lesson of the Learning Event (and more generally) is to begin to work to create a process, and to subordinate design remade to it, recognising that: critical and appropriate questions can be discovered in time; problems can be rigorously defined in relation to their locus in the unsustainable (in its relational complexity rather than as a biophysical reduction); and, the search for solutions, or means of adaption, is predicated upon Sustainment as the imperative to which to respond.

1   See Sustainment entry at www.thestudioattheedgeoftheworld.com.
2   This is an extract from an unpublished 'The Studio and the Edge of the World' Learning Event briefing document.

References

Fry, T. (2010). *Design as Politics*. London: Berg Publishers.
Fry, T. (2008). *Design Futuring: Sustainability, Ethics and New Practice*. London: Berg Publishers.
Macpherson, C.B. (1973). *Democracy Theory: Essays in Retrieval*. Oxford: Clarendon Press.
Fry, T. 'The Studio at the Edge of the World'. Unpublished essay, Learning Event briefing document.
Sustainment entry at www.thestudioattheedgeoftheworld.com.

RESEARCH FOR DESIGNING AFTER OWNERSHIP

Cameron Tonkinwise

A Macroeconomic History of Design

The phases of capitalist modernisation, in which design has played a crucial role, could, very crudely, be characterised as:

- from people doing skilled things with manual tools themselves;
- through people getting other people to do skilled things with manual tools for them;
- to people doing less skilled things with powered 'labour-saving' devices that they own.

This schema – roughly adapted from Walter Stahel's insight that the twentieth century involved the replacement of people with (fossil-fuelled) devices, something that should now be reversed in the name of sustainability (Stahel 1981) – describes the shift in (North Atlantic) households from village life, through bourgeois servant-based society, to suburbia. The latter can only entail small nuclear families living more or less autonomously in 'free-standing' dwellings if the storage spaces of those buildings – the closets, attics, basements, and garages – are filled with 'labor-saving devices'.

The professional practice of modern design – by which I mean mostly industrial and product design – flourished as a result of this shift, especially from the second paradigm to the third. Designers were called upon to give form to technologies so that those devices could be domesticated into the products people could use instead of outsourcing service provision to slaves or servants or service businesses. While designers have always been involved in giving form to technologies employed in the workplace, designing came into its own as a modern profession in relation to consumer goods, shaping what and how technologies arrived into households throughout the twentieth century.

This expertise in product styling that developed into consulting design expertise in the pre-Second World War period achieved pervasive agency with suburbanisation in the Global North throughout the second half of the twentieth century. New land-use strategies required new types of buildings, each of which needed to be filled with the conveniences of modern, independent living – and designers were called upon to make decisions about how those home conveniences should look and function.

I give this crude macroeconomic history of modern design because it fore-grounds the intertwining of design and household ownership. Design mediates be-tween the constraints of mass production and the variables of mass consumption, transforming technologies into goods that prospective users want to own. Owning has been the central working assumption of design for the last century. This had at least three consequences for the nature of design research until recently that I would like to underline.

Visibly Distinct

Design is first if not foremost about attracting the attention of buyers. If designing is primarily about getting a design owned, the designer must make the product no-ticeable and then desirable. As we know from Everett Rogers' *Diffusion of Innovations* (2003), new products must also be perceptibly useful and trialably usable, but these are subsequent to the product first breaking into the awareness of a prospec-tive buyer. This is why the modern process of designing (consumer goods) drives primarily at novelty. What is being designed must clearly differentiate itself from what currently exists. And this then is, or rather was, one of the central objectives of design research.

As we know since Herbert Simon (1996), researching what else might be, rather than what merely already is, pursuing precisely what is not yet the case, finding new-ness, is a methodological challenge. It is what led John Chris Jones to propose that the ultimate design method was randomisation – see his preface to the second edi-tion of *Design Methods* (1992) and his collection of essays *Designing Designing* (1991). When conventional research undertakes 'gap analysis', it must also be sure that what is missing is significant. By contrast, the significance of the new remains for design to make rather than have surety about beforehand. When researching novelty, any-thing is constructivistically more or less possible, so methods must be generative.

But more important than the privileging of novelty, given design's focus on ownership, is the centrality of the visual. Since designers must get designs to claim the attention of shoppers, what is new about those designs must be strikingly visible. This has meant that design research over the twentieth century has been strongly ocularcentric: colour theory, shape grammar, product semantics. This is not only the case for these domains of 'research for designing' but also in the more academic 'research of designs', that is to say the more material culture side of cultural studies, where visual semiotic readings of designs dominate.

Petran Kockelkoren and Peter-Paul Verbeek (1998) pointed out that this has a paradoxical effect for both the designer and the owner/user. Following Platonic logic, the actual material product appears to be the poor substitute for the imagined ideal that the product's visual design signals or aspires to embody. Jean Baudrillard

(1996) points out that the resulting designs tend to become 'signs of themselves', and so subject to serial replacement by both the designer trying to pursue an iconic form, and the consumer never getting what they thought was promised.

Identifiable Style

A second related consequence to designing for ownership concerns the class politics of taste. The birth of design was related to the democratisation of goods made possible by cheap mass production. This resulted in a transformation of class distinctions. Consider for instance celluloid collars, which allowed people who could not afford servants to starch their collars each day to dress as if they did have collar-starching servants. Designers responded to this mixing up of class identifiers by further proliferating product forms. The result, as Pierre Bourdieu has outlined (1984), was the opportunity for groups within financially equal classes to differentiate themselves from each other in terms of cultural and social capital through their choices from that variety of product forms. The same kind of product could be designed with different forms to fit with the distinctive taste of different classes within society.

The primary task of design research then becomes that of identifying class tastes. People's aesthetic preferences are carefully examined so that the form of the product can blend with their overall (life)style. Branding moves from being about the producer communicating their values consistently across their product range, to a matching of product forms to the semiotic signals of consumer segments. Even when design begins to pay more attention to usability as part of the value differentiator of design, the focus on visible class markers continues to remain central to design research. I am thinking here of personas (Cooper 1999): target market demographics and psychographics are synthesised by design teams into visualised mood boards of the aesthetic preferences of two or three fictional individuals, who are then 'consulted' as to the most appropriate forms of interaction for what is being designed.

The crucial point here is that these marketing-oriented design-sociologies of primarily urban consumer society presume ownership. Possessions are considered part of what Russell Belk (1988) calls 'the extended self'. The job of a designer is to allow people to express themselves through the objects they possess and so dwell with each day. These products, or rather their visibly broadcast forms, display the identity, or identify the class, of their owner to others who might visit the owner's property or see the owner wearing them down the street; but they also broadcast their style back to the owner him- or herself, reinforcing the identity the owner thinks of themselves as having. In some cases, products do the work of communicating class identity by being owned even without being used: the aspirational musical instrument in my living room, the no longer used sporting equipment in my garage, the just-in-case formal wear in my wardrobe, the unread books on my bedside table; each declares not just what I actually am, but what I could be, if only such occasions arose (again).

There is the possibility that I can get the class benefits of design without ownership, like when I am seen driving a sports car that I have in fact rented; but if that identification happens, it does so because there is an assumption, in that case mistaken, that I in fact do own that designed product, that it is therefore part of my own (extended) self. The designer's job is therefore to research the meanings that will be felt by someone being seen with such a possession.

Possessing Use

A third consequence of ownership-oriented designing is that designed products are able to involve considerable 'learning curves'. If products are to be owned for a significant period of time, then there is both the motivation and the time to learn to use them. This may explain why 'user-centred design' arose as an innovation in the design process more than fifty years after design had arisen as a profession. One can ask, what was design before user-centred designing? While it was 'designing for people' to quote the title of Dreyfuss' 1955 text (2003), this designing was more about physically limiting designs to fit (universalised) human factors, so that products were minimally capable of being used, rather than researching new ways of making products easier for distinct sets of people to use. The designer of Dreyfuss' era would research the task that the product facilitated, but only with a view to generic human parameters. This was because a period of learning to use the owned product was to be expected. Owners could, and so should, take the time to read manuals and train themselves in the use of designed goods.

In some ways, it took the politics of Scandinavian unionists to focus designers on making tools easier to use, to reduce the expertise and/or training time required of workers (Ehn 1988). This became the discourse of human-computer interaction, paving the way for the user-centred design movements associated primarily with the domestication of the computer in North America. However, before this politically motivated prioritisation of usability, the designer's job was merely to ensure functionality rather than immediate 'out of the package' usability.

Putting these tolerated 'learning curves' together with the previous point about the visual semiotics of design may explain why designs within an 'ownership' society approach functionality as a process of decoding. Design prior to the user-centred design movement involved mostly the application of signs to products. Owners of newly designed products would train themselves in a series of commands and processes involving those designed symbols. This is quite distinct from the current, more enactive approaches to usability: these aim for a minimum number of interactions each of which manifests more intuitively to users through affordances.

Importantly, it was not just that economic ownership allowed people to invest in learning how to use products, but also that this learning time enhances the psychological aspects of ownership. Kaj Ilmonen (2004) has given a comprehensive account of the way in which product learning curves entail psychical investment in those products. Feeling an attachment to a product is also a result of having spent a period mastering the expert uses of a product, incorporating it into both your daily life and your body. That sense of control is a strongly affective form of property, but it only develops because the product is from the start owned by the user who progressively skills up with the product until it is a habitual part of their habitat.

There was the assumption that this economic and psychical sense of owning things would lead owners to value their things, maximising the longevity of their productive use. However, this proves in most cases not to be the case. Ownership, for most categories of product, is invariably inefficient, at least in terms of time-budgets, as that product sits idle in household storage awaiting very occasional use. Caring about a product, as a result of money invested in buying that product and time invested in learning to use that product, frequently does not translate into the quite distinct practice of actively caring for a product (Howarth 1996) – so that when it occasionally needs to be pulled out of storage, it sometimes fails due to lack of maintenance. Though designed for ownership, products for the most part are not in the end productively owned.

The Coming Age of Post-Ownership

For almost half a century, there has been talk about a radical shift in design, from industrial to post-industrial (Balcioglu 1998, Cross 1981). I want to suggest that over the last decade something like this shift has finally occurred. But it has not been the predicted shift away from physical product manufacturing. The very information and knowledge economies that were thought to characterise the post-industrial era in fact depend on a proliferation of digital devices (Moles 1988). Much hoped for convergence towards fewer devices seems to be continually countered by accessory productisation, but more so by accelerating series releases. The design of hardware, despite being physicalisations of energy-intense rare earth metals and fossil-fuel based polymers, seems to be infected by the same 'iterativism' that is more afforded by the software they contain. In short, product-centred industrial design has not been displaced by interaction design. What has in fact changed in respect to 'post-industrialisation' is the nature of the ownership of those products.

Accessing Other People's Things

The most evident instance of this recent shift has been the rise of what is now called the 'Sharing Economy'. The title is certainly a misnomer as it covers a wide range of economic relations.

The most direct version is peer-to-peer exchange of goods. The ways in which the internet has enabled search (see Anderson's notion of 'the long tail', 2008) and lowered transaction costs (including those associated with trustworthiness as a result of 'true identity') quickly facilitated secondary markets for the resale of owned goods (e.g., eBay). Related was the possibility of peer-to-peer lending or leasing of goods.

In the mid-2000s, there was the hope that these kinds of business models – working alongside systems of pooled-goods owned by businesses but available for on-site use on pay-per-use basis (e.g., launderettes, print shops, etc.) – would significantly reduce societal materials intensity (Tukker and Tischner 2006). Performance Economies (Stahel 2010) centred around functional equivalence – 'you don't want (to own) a washing machine or drill, you want (access to) clean clothes or a hole' – would encourage people to make more productive use of fewer things. Products would have multiple users, either by moving through a series of owners (redistributive markets), or by being owned by a business selling the use of the goods (fleet management), or by being owned by an individual or household providing other individuals or households temporary access to the goods (peer-to-peer lending). These research-led attempts at 'sustainable servicisation' proved a few years premature (Tukker 2004).

When the subsequent rise of social media platforms better afforded these scenarios of 'usership', the focus of such 'product service systems' shifted towards the sharing of labour as opposed to the sharing of goods. The most prominent platforms in the 'Sharing Economy' today are Uber and Airbnb. In either case, what is being exchanged is a privately owned asset along with service provision by the owner: a participant is 'borrowing' (with compensation) someone's car as well as that someone, as the driver; or a room plus that landlord's hosting (to varying degrees). In some ways, this looks like an unchanged form of ownership: the owner of the private asset being shared remains the user rather than handing over complete control to the recipient of the sharing economy service. The difference however is that the owner of the private goods is prompted by being a service provider in a Sharing Economy platform to retrofit their goods so that they better suit a range of users; a car must be kept clean with water for riders; a room must be redecorated or even furnished with a less personal identity for visitors. An Uber driver may legally own their car, but use of it for ride-sharing will make it look and feel less personalised, more like something other people use; the room in the property I own that I let on Airbnb is not a room that I ever go into, except to service it for guests. These systems for on-demand labour provision therefore also involve a shift away from the 'ownership-oriented' design's focus on private household-centred lifestyles through-out the twentieth century (in the consumer capitalism of the Global North). Design-ers are just now starting to design interiors of cars and houses that will be used for Sharing rather than for conventional forms or owner-occupiers.

Unstable Digital Landscapes

Less apparent than these 'Sharing Economies', though perhaps more pervasive, are different models of economic possession associated with digital goods. Firstly, rapid developments in computational capacity, and so, in parallel, rapid develop-ments in systems using that increasing capacity, have normalised a high turnover of information communication technology products by the Global Consumer Class. The constant updating of software and then hardware that began with desktop PCs needing to be repurchased every two to three years, has now transferred, in an ac-celerated manner, through mobile phones to the increasing range of products with information technology components – not only wearables but even kitchen appli-ances and sporting equipment. As a result, shortening periods of ownership are re-sulting in purchasing agreements that are closer to leasing. Few people own the routers upon which their household leisure, but also more and more their work-from-home capacity, depends; just as most mobile phones are paid for via subscrip-tion models with built-in upgrade periods.

When sustainable designers were petitioning for extended producer respon-sibility in the 2000s, they advocated a shift to Leasing Economies as a way of in-centivising Closed-Loop Systems (Cooper 2010). The current Tech Economy is a de facto Leasing Economy though without much of the ecologically responsible prod-uct take-back for component recovery.

But it is not only the economic relations behind digital device use that are less like ownership. This also manifests, secondly, in the more psychical relations in-volved with digital devices. Consider that there has been a perhaps ontological shift in the nature of interaction as a result of Apple hand-held devices – though Apple was not the sole cause, just its most prominent capitaliser. Before the touch-screens of iPods and iPhones, things that attracted interactions were mostly fixed physical forms: switches, buttons, handles, etc. – i.e., the visible signs of interaction I dis-cussed in 'Possessing Use' earlier. We now, especially after the abandonment of skeuomorphism, can expect interactibility from any kind of finger-on-glass move-ment – e.g., swiping to scroll without a scroll-bar being present.

The obverse of this is the capacity for different applications to reconfigure the screen from moment to moment for completely varied interaction experiences. Con-sequently, while I may more or less own my 'personal digital devices', my capacity to feel like I have controlled ownership, in the sense that Kaj Ilmonen describes as a kind of mastery, is limited. Each new app installed on my devices, or even each new update of currently installed apps (something that increasingly happens automatically in the background beyond my control), reconfigures the nature of the designed artefacts I thought I possessed.

Since the interface is modifiable by the manufacturer/designer, what is owned by the user is the data: the collection of apps and their arrangement; the photos and social media posts; the emails and downloaded music, videos and books; etc. But these mostly do not reside in the device, or at least only in the device. They exist rather in the 'cloud', an abstraction for which users must invent imaginaries of 'device landscapes' (Stolterman et al. 2013). A consequence is that if this particular physical manifestation of the device the user 'owns' is dropped or lost, the user can buy another that is physically identical, and almost immediately restore what is 'theirs' about the device, even though this is a distinct, new product. As Karl Jaspers identified at the start of the twentieth century, mass production of the identical means that there is nothing inherently valuable about 'this one' (cited in Verbeek and Kockelkoren 1998). This is now much more the case, when that 'oneness' comprises digital copies of files that can appear identically on a wide range of devices.

And of course, while I 'own' my data, we are becoming more aware that the price of a more or less 'free' internet is loss of ownership of my '(meta)data'. What I do with my files, as well as traces of all my physical actions, such as my location, while using those files, is being collected and sold to systems of surveillance capitalism.

This then is what I believe to be the real shift that design research must now take into account. The devices and systems that characterise current post-industrialism in fact still involve significant aspects of industrial design. What has transformed radically is the context in which those industrially designed post-industrial products are used. That context is no longer based on the system of individual, or at least household, ownership that was structurally determining of how (industrial) design conducted its research-led practice. Designers must now design products and the ways in which people interact with them in situations of complicated, and transforming, not-quite-ownership. To explain what this involves, I will review the points made so far.

From Attract to Persuade

There is still a substantial market in goods bought through the practice of shopping for conspicuous, semi-durable goods to own. But those physical goods, especially the range of digital devices, tend these days to be difficult to distinguish in product form without close inspection. Ongoing convergence and miniaturisation allow certain product categories to shift into less obtrusive wearables, pervasive computing or household infrastructures. Applications purchased for those goods are also subject to increasingly comprehensive aesthetic constraints (e.g., Google's *Material Design* guide: https://material.google.com), in part to control the experience across varied platforms (what is now called Responsive Design), but also to maintain the platform's brand identity across a number of outsourced developers. In other words, precisely because the value proposition of these economies is distinct from the perhaps owned physical product, visual distinctiveness is deprioritised.

As a result, designers must find other ways to enlist users in innovations. These tend to be at either end of a political spectrum: participation or capture. The research done by designers in technology companies tends to occur within Lean Startup / Lean UX frameworks at the moment. These involve action research through Minimum Viable Product releases to lead-users. While prospective users are consulted in the innovation process, these processes should be differentiated from the more politically motivated practice of participatory design. Though more than just user testing, the experimentalism is mostly too centred around 'engagement' with the product, platform, or application (see metrics like Daily Active User) to give users the capacity to raise concerns outside the sought-after value-proposition. The point for now is that research for these kinds of usership, as opposed to ownership, takes places more as 'by design' than 'for design'.

The reverse forms of research are systems that aim to be 'sticky' for users, or nudge users into subscriptions. Precisely because the designer is not looking for the comprehensive intention involved in owning, but merely for user engagement with a product-service, the research is less concerned with intentions and desires, and more with implicit biases and only ever semi-conscious habits. Data traces of actions indicate how to position products in places that users adopt without being fully aware of meaning to. To put it more politically, when design concerned product use within systems of ownership, design's functionalism was tempered by other concerns such as aesthetics; usership-oriented design research is now liberated to be hyperfunctionalist.

From Display to Perform

Cultural critics have long decried consumerist materialism for its inauthenticity and rivalrous sociality. There is the hope that the shift to post-industrial usership might afford a society that is more centred on 'doing' if not 'being' rather than 'having'. The academic trajectory of Joseph Pine and James Gilmore provides an example of this transition – from *Mass Customization* (1999), through *The Experience Economy* (1999) to *Authenticity* (2007) – as do claims about the Access Economy (Rifkin 2000) and *The Support Economy* (Zuboff and Maxmin 2004).

Certainly, design research is required to now be more 'activity-centred'. Designers must attend more to verbs than nouns. If products are present, they are 'service carriers', just one component of constellations of a Social Practice. Design research then turns away from the visual style of this or that product's form to the more embodied style of how an experience unfolds over time. Style now refers more to the feel of an interaction, the tempo and rhythm of a service journey.

However, both these tendencies are somewhat countered by the pervasiveness of social media. These reassert the ocularcentrism of ownership-centred industrial design as people seek appreciation for their self-displays.

From Learn to Co-Evolve

As product possession diminishes or is subject to constant digital modification, designers can no longer assume users' tolerance for learning interaction symbols and steps. This increases pressure on designers to deliver immediately usable designs to the market. Designing must be much more directed by a comprehensive and intimate understanding of users' current patterns of interaction.

There is a risk that these imperatives lead to a design conservatism, both across cultures and over time. Some have started to complain about a ubiquitous 'modern universal' aesthetic in product forms and interfaces (Michl 2014) – and certainly this is evident in the monopolistic technology corporations' 'Design Guides'. Innovation is possible but it must just be more incremental than in the previous ownership-based regime. As noted previously, design research needs to be directed towards sociotechnical co-evolution: users can prompt, or be prompted to make, a minor shift in a product series' or platform's modes of interactions, that then makes possible a subsequent new practice, which can in turn be amplified by another redesign. This appears to have been Apple's decade-long strategy for introducing iPods, then iPhones, then iTunes-based cloud-computing, then iPads, and now watches (perhaps too great an interaction 'evolution') and soon conversational user interfaces through wireless earphones, etc.

In addition to conservatism, there is a danger that design's over-emphasis on immediate usability results in a form of de-skilling. Devices are blackboxed behind intuitive interfaces, and automation and even predictive systems are prioritised, as design aims to make everything as frictionless as possible. This applies not only to interfaces for existing systems, but to the kinds of problems that design research prioritises for innovation: how to enable meal preparation without learning how to cook, how to be mobile without learning how to drive or even how to make use of a public transport system, how to learn at school or university or in the workplace by playing games that do not involve learning how to interact with people or texts or concepts that are difficult.

Many suggest that the emergence of the Sharing Economy signals a return of general trust (Sundararajan 2016). There are several reasons to think this is not so much the case. For a start, if the arrival of post-industrialism correlates with a reduced sense of ownership, then people need not trust as much when lending or exchanging goods, because any prospective loss is reduced. Trust is perhaps more necessary when the Sharing Economy involves peer-to-peer labour provision, but two other factors then come into play.

The first is that social media appear to represent not just a migration of pre-digital forms of sociality onto online platforms. Rather they involve new kinds of sociality. The verb 'to friend' signals relations to people who are neither friends nor strangers. These kinds of weak ties can be operationalised because of the search mechanisms and network effects of the internet. When engaged in Sharing *Economies*, 'peers' are neither part of a shared community nor merely economic agents concealed behind the role of customer and employee. Moving outside formal legal commercial contracts does not necessarily locate those exchanges in the realm of personal trust. The latter is more accurately the realm of identity and property, exactly what I believe is loosening. The task of the contemporary design researcher must therefore be to remain open to these new kinds of socially embedded economic relations, noticing new kinds of sociality afforded by mobile connectivity. To put it more pointedly, the design researcher is not looking for existing identities, but emerging shifts in the nature, or rather practice, of cultural identity.

The second factor is that many of the interactions that fall under what is being called the 'Sharing Economy' are being performed under particular constraints. Foremost is economic restraint following the Global Financial Crisis. This could be a direct impact of losing employment or stagnating wages, or an indirect impact of subsequent neoliberal austerity measures that erode government services. If I begin to participate in the Sharing Economy because I feel economically compelled – to use a cheaper service or to try to generate revenue by offering such a service – issues of trust will be significantly tempered. At the other end of the economic spectrum are those whose productivity demands participation in aspects of the Sharing Economy where usership is more convenient than ownership (if some of the costs associated with ownership and service provision are being externalised onto the former economically restrained service providers). In either case, in line with what I have argued previously, the focus of the design researcher must be more on the experience of the practice than the apparent meaning of the practice. If the system of shared use is efficient and effective, then that functionalism will override more industrialist concerns about property.

The ubiquity of information and communication technology that is part of post-ownership-industrialism was predicted to have rejuvenated the Enlightenment (Beck 1995). Instead, barbarism with respect to expert knowledge and otherness seems ascendant. In conditions of liquidity (Bauman 2000), without forms of community that are associated with settled modes of dwelling with more static possessions, people seem to have embraced other forms of communities of interest that can be self-reinforcing in terms of scepticism and xenophobia (Tonkinwise 2016).

There is therefore a responsibility for design research to find ways of re-establishing the capacity for people to engage in a considered manner with the risks associated with current ways of organising society and emerging futures. Some of this appears to be happening when design researchers engage in Critical Speculative Design. These para-functional explorations are undertaken to provoke debate about prospective futures. In most cases, the designers do not take responsibility for ensuring that those debates in fact take place in ways that allow communities to make decisions about such futures, but that kind of research certainly could and should.

Conclusion

The post-industrialist design researcher, given movements away from twentieth-century notions of owning designed products, must now focus on social practices and the ways in which new kinds of socialities are afforded by those social practices. This requires designers not just to be much more anthropological and sociological than they needed to be when imposing goods for purchase on suburbia, but also to be more political. Research takes place not only before designing, or even by design, but also after release of the design into the market. This enables designers to take more responsibility for the socially complex consequences of what they are designing. They can and must now 'stay with the trouble' precisely because what they are designing no longer disappears into suburban households as finished possessions (Tonkinwise 2005).

References

Anderson, C. (2008). *The Long Tail: Why the Future of Business is Selling Less of More*. New York: Hyperion.
Blacioglu, T. (ed.) (1998). *The Role of Product Design in Post-Industrial Society*. Ankara: Middle East Technical University Press.
Baudrillard, J. (1996). *The System of Objects*. London: Verso.
Bauman, Z. (2000). *Liquid Modernity*. Cambridge: Polity.
Beck, U. (1995). 'Ecological Enlightenment: Essays on the Politics of the Risk Society'. *Organization & Environment*, 11(1).
Belk, R.W. (1988). 'Possessions and the Extended Self'. *Journal of Consumer Research*, 15(2).
Bourdieu, P. (1984). *Distinction: A Social Critique of the Judgement of Taste*. Cambridge, MA: Harvard University Press.
Cooper, A. (1999). *The Inmates are Running the Asylum*. Indianapolis, IN: Macmillan.
Cooper, T. (ed.) (2010). *Longer Lasting Products: Alternatives to the Throwaway Society*. Swansea: Gower Publishing.
Cross, N. (1981). 'The Coming of Post-Industrial Design'. *Design Studies* 2.1.
Dreyfuss, H. (2003). *Designing for People*. New York: Skyhorse Publishing.

Ehn, P. (1988). *Work-Oriented Design of Computer Artifacts*. Stockholm: Arbetslivscentrum.

Howarth, J. (1996). 'Neither Use nor Ornament: A Consumers' Guide to Care'. *Thingmount working paper series on the philosophy of conservation*.

Ilmonen, K. (2004). 'The Use of and Commitment to Goods'. *Journal of Consumer Culture* 4.1.

Moles, A. (1988). 'Design and Immateriality: What of It in a Post Industrial Society?' *Design issues* 4, no. 1/2.

Jones, J.C. (1991). *Designing Designing*. London: Architecture Design & Technology Press.

Jones, J.C. (1992). *Design Methods*. Hoboken NJ: John Wiley & Sons.

Michl, J. (2014). 'Taking Down the Bauhaus Wall: Towards Living Design History as a Tool for Better Design'. *The Design Journal* 17.3.

Pine, B.J. (1999). *Mass Customization*. Boston, MA: Harvard Business Press.

Pine, B.J., Gilmore, J.H. (1999). *The Experience Economy: Work is Theatre & Every Business a Stage*. Boston, MA: Harvard Business Press.

Pine, B.J., Gilmore, J.H. (2007). *Authenticity: What Consumers Really Want*. Boston, MA: Harvard Business School Press.

Rifkin, J. (2000). *The Age of Access*. New York: Tarcher Perigee.

Rogers, E.M. (2003). *Diffusion of Innovations*. New York: Simon & Schuster.

Simon, H.A. (1996). *The Sciences of the Artificial*. Cambridge, MA: MIT Press.

Stahel, W.R., Reday-Mulvey, G. (1981). *Jobs for Tomorrow: The Potential for Substituting Manpower for Energy*. New York: Vantage Press.

Stahel, W. (2010). *The Performance Economy*. Basingstoke: Palgrave Macmillan.

Stolterman, E., Jung, H., Will, R. (2013). 'Device Landscapes: A New Challenge to Interaction Design and HCI Research'. *Archives of Design Research* 26.

Sundararajan, A. (2016). *The Sharing Economy*. Cambridge, MA: MIT Press.

Tonkinwise, C. (2005). 'Is Design Finished? Dematerialisation and Changing Things'. *Design Philosophy Papers* 3.2.

Tonkinwise, C. (2016). 'Designing in an Era of Xenophobia'. *The Radical Designist* 6.

Tukker, A. (2004). 'Eight Types of Product–Service System: Eight Ways to Sustainability? Experiences from SusProNet'. *Business Strategy and the Environment* 13.4.

Tukker, A., Tischner, U. (eds.) (2006). *New Business for Old Europe: Product-Service Development, Competitiveness and Sustainability*. Yorkshire: Greenleaf Publications.

Verbeek, P., Kockelkoren, P. (1998). 'The Things that Matter'. *Design Issues* 14.3.

Zuboff, S., Maxmin, J. (2004). *The Support Economy: Why Corporations are Failing Individuals and the Next Episode of Capitalism*. New York: Penguin Random House.

DESIGN AND OPEN INNOVATION FOR SUSTAINABILITY: LET'S GET RADICAL

Ursula Tischner

This chapter summarises the rationale behind why we need radical Design and Open Innovation for Sustainability, the basic principles, and the role of research in that context. It will respond to questions that are frequently asked by designers and other creatives such as, why should designers get involved in saving the world, and is it really so urgent? It will further discuss some of the emerging and exciting new opportunities to step out of the predominant way and understanding of doing design to show that, even though the mainstream economic system still is the wrong framework for true sustainable development, there are many alternative pathways to follow. Conclusions are drawn for a more radical and innovative understanding of the role of design and designers fostering change and transition towards a more sustainable way of living.

A Fundamental Question: (Why) Should Designers Save the World?

The answer is: Well, why should they NOT? It makes sense to ask all professions to think about and take into account what they can contribute to making the world a better place for as many people as possible, avoiding negative social and environmental impacts as much as they can. Especially politicians, managers and decision-makers in companies, educators and opinion leaders, investors and lobbyists etc. have a great deal of responsibility here. But designers, architects, creative engineers among other creative professionals are really in a pole position when it comes to inventing new goods and ways to organise daily life and work routines and to 'sell' these novelties to consumers and clients – that is the core of the design profession. Designers can even encourage new lifestyles and social innovation. With these capabilities they can do really negative things but also really positive things in terms of sustainability.

There is much evidence that in the past and also presently designers are more part of the problem than part of the solution. They still act in the middle of the consumerist society to spur on mindless consumption using up our natural resources at an alarming rate. For a historical view Victor Papanek's books *Design for the Real World* and *The Green Imperative* are recommended (Papanek 1985, 1995). Also Vance Packard described clearly in his publication *The Waste Makers* the invention of planned (built-in and perceived) obsolescence by designers, marketeers, and managers after the Second World War to promote consumption for consumption's sake (Packard 1960). Packard detected back in 1960 that business was making us 'more wasteful, imprudent, and carefree in our consuming habits'. This phenomenon has become much worse today. In fast fashion industry, for instance, fashion designers create twelve to eighteen or even more 'seasons' per year and the time span from the sketch of the designer to the garment in the retail store is as little as twelve to fourteen days (van Markoviec 2009).

There is a notorious rule of thumb that during the product development and design phase about 80 per cent of the overall environmental (and social) impacts of products are determined (as are the economic aspects). That is the reason why this phase is so important for making a product or service more or less sustainable and this responsibility lies with the designers and the product developers as well as the strategic planners writing the design briefs, and their bosses.

Unfortunately, many practising designers do not have the right knowledge and do not use tools that help them to evaluate the environmental and social impacts of their designs and to integrate Sustainability thinking into their daily work routines and design processes. Perhaps this is the reason why so many sniff at Sustainability Design or even get angry when confronted with it. Judged with thorough sustainability criteria, much of the design we see on the market today would not rank very highly.

This unfortunate ignorance is still prevalent despite the fact that there are so many helpful tools for Design for Sustainability and Ecodesign, many practical guidelines and extremely good information portals, legislation, and ISO standards etc. Still, when asking designers why they are doing so little, these are their main arguments:

- We do not have the relevant information at hand.
- We do not know how to judge whether something is more or less sustainable, cannot know what sustainability really is (and no one really knows).
- We are too busy to deal with this.
- No one asks us for this and no one pays us for it.

So, to respond to the first two issues, the author of this chapter has already written and edited two books: first, *How to do Ecodesign* (English version) / *Was ist Ecodesign* (German version) in 2000, commissioned by the UBA (Umweltbundesamt), the German version of the EPA, which was completely updated in 2015 and is available free in its second edition as iBooks and e-books. This practical guide includes the Eco-/ Sustainable Design Toolbox with many helpful methods and tools, and explains how and where to use them in the design process for which purposes.

The second publication entitled *Changing Paradigms, Designing for a Sustainable Future,* edited by Peter Stebbing of the Hochschule für Gestaltung in

Schwäbisch Gmünd, and the author of this chapter for the Cumulus International Association of Universities and Colleges of Art, Design and Media, is especially meant for those who are interested in research about the 'why' we need design for sustainability and the 'what', and 'how' to do it.

It also gives many student project examples from Cumulus member schools.

This leads to the other big obstacle working against Sustainability Design becoming mainstream: the lack of educational programmes that offer Sustainability knowledge to design students and professionals. There are a few, but Sustainability is not yet a standard subject in design education, as for instance ergonomics (human factors) is.

Tackling the last two arguments above – lack of time, demand by clients and payment – it can be reported that there is growing demand for Sustainability Design expertise. That is the reason why most of the large design agencies work in the field of Eco- and Sustainable Design, the big consulting companies offer Ecodesign consulting, and more and more companies seek expertise in this field and create positions, such as Sustainable Design, etc. Especially under the new catchword 'Circular Economy' (which is actually not that new, as countries such as Germany have had a circular economy law since September 1994), growing interest of industry emerges. It seems that including the word 'economy' already sounds more promising for profit-oriented actors. Accordingly, green job portals list many sustainability-related positions. Research by *WIRED* magazine in the year 2012 showed that 'green' and sustainability jobs were the only job category with over 50 per cent growth rates in the previous five years in the US. But designers will only be able to do a good job in this field and thus be sufficiently paid for it, when they take the time and effort to acquire the necessary knowledge.

Another Fundamental Question: Is it Really so Urgent?

The author has witnessed many discussions among designers about legitimacy and urgency of designing, for instance, products like baby nappies with sensors that tell parents when they need changing, versus designing solutions confronting climate change, resource depletion, water scarcity, and the like. Designers often seem to think that the planet is still in a good state and they can continue the way they have designed in the past for the rest of their career. They lack the sense of urgency and the motivation to change their understanding of their role as a designer and the way they approach designing.

Thus it might be helpful to discuss some facts about the state of our planet below.

It is important to realise that we talk about the physical state of the planet, as it is described by in-depth research, and the impacts human physical activities have on it. This is nothing we can interpret or negotiate as designers. From all the research that has been done by eminent scientists it is pretty obvious that the way and speed we humans use resources for our purposes, change them and give them back to the planet as emissions, effluents, waste, and other pollutants, will get us in deep trouble fairly soon. Actually it has already. Why do we have a 'refugee crisis' in Europe? And what are the real reasons for most current conflicts and wars? People are fighting for access to water, resources, fertile land, opportunities to make a living etc. And what will happen, if climate change makes things worse?

The following three research activities/groups will be used to illustrate this point:

(a) **The Ecological Footprint calculations by Mathis Wackernagel and the team of the Global Footprint Network** who try to assess the ecosystem services that our planet offers and compare these to what humans consume living on this planet per country and overall.[1]
Among other results of these kinds of calculations are that

- if everybody wanted to live like the average North Americans do, we would need six planets,
- if everybody wanted to live like average Germans do, we would need four planets, and
- the so-called Earth Overshoot Day (when we have consumed all ecosystem services that the planet can renew within one year) happens earlier each year, in 2017 it was by 2 August.[2]

From these calculations, the conclusion can be drawn that it is an urgent goal to reduce the absolute resource (including water, energy, land etc.) use of people in industrialised countries to allow so-called developing and emerging countries to increase their level of consumption and quality of life, especially for the people living below the poverty line. For Germany that means that we would need to reduce the current consumption of resources by a factor between 4 and 20, depending on the resource we are looking at, by 2030 and 2050, respectively (KRU 2014). So designers could ask, how can we design 'stuff', services, systems that work towards this kind of 'dematerialisation' without losing or even with increasing quality of life.

(b) **Johan Rockström and his team's work on the so-called Planetary Boundaries**.[3] According to their research, humanity has already exceeded three or four of the ten boundaries of our planet. Especially loss of biodiversity, the nitrogen cycle, the phosphorus cycle (very important for agriculture, i.e. food production) and of course climate change have already exceeded the limits.[4]

(c) And finally **climate change**. There is no longer any need for a debate on whether climate change exists and whether it is (partly) created by human activities. Since the COP 21 meeting in the year 2015 in Paris, the world has agreed on a common goal, which is not up for negotiation any more. The tangible challenge here, that can also be tackled in design, is to limit the average temperature rise of the planet to under 2° Celsius (better 1.5° C) compared to pre-industrialisation levels. That means that we have to limit the concentration of CO_2 and other greenhouse gases in the atmosphere to 450 ppm. In the year 2015 the concentration was already 485 ppm and in the record-high year 2016 it was 489 ppm (NOAA). Thus the world community urgently has to strive for an absolute reduction of global greenhouse gas emissions. Many experts agree that the next ten to fifteen years are decisive to take action before the point of no return, the 'tipping point', is reached (UNEP Report). Here designers could question, how their design activities contribute to that goal, again, without losing or even with improving quality of life for as many people as possible at the same time. Important fields for action are: eliminating the use of fossil energy sources, transport and mobility, agriculture and eating less meat, reducing deforestation, increasing energy efficiency, taking care of biological waste (methane emissions) etc.

All this evidence asks for a radical change in the way we design products, services, systems, as well as communication and education towards a new design paradigm, which we can call 'Design for Sustainability'.

Design for Sustainability

The terms Sustainable Design, Sustainability Design or Design for Sustainability are closely linked to the vision of Sustainable Development. Sustainable Development, as originally defined by the United Nations' (UN) Brundtland Commission (WCED 1987), aims to provide the necessary environmental, social, and economic

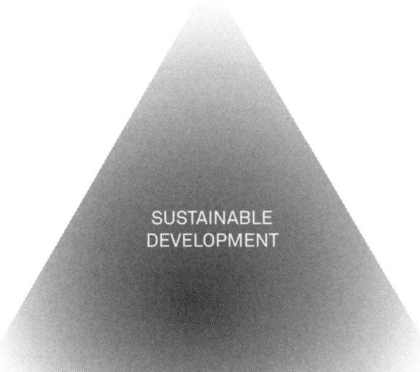

1 The three pillars of
Sustainable Development
(original definition)
(Tischner et al. 2015)

PROFIT (economy)
- stable economy development
- growth of welfare
- employment
- affordable goods
- balance between countries

SUSTAINABLE
DEVELOPMENT

PLANET (ecology)
- respecting the carrying capacity
  of nature
- sustainable use of renewables
- minimal use of non-renewables
- closed-loop economy

PEOPLE (social-ethical aspects)
equal distribution of opportunities
- between people, men and women
- between poorer and richer
  countries
- between the generations

conditions to allow for the survival of a growing world population on a planet with limited resources. Key criteria and measures to achieve this are proposed in Agenda 21 (UN 1992). This definition suggests that it is necessary to aim at environmental, social, and economic improvements simultaneously. In addition to the three pillars of sustainability (people = social, planet = environmental, profit = economic), a fourth is often added at the institutional/policy implementation level.[5] As this approach did not emphasise strongly enough that the social, economic, and institutional levels cannot exist, if the natural environment is destroyed, the more recent Sustainability 2.0 Paradigm has been developed that sees the environment as the basis for all other levels.

Design for Sustainability (DfS) thus produces solutions that provide meaningful benefits to society, improve quality of life (in particular for less wealthy people), create added value for suppliers and customers, while avoiding damage to or even positively interacting with the natural environment.

Ecodesign (often used as a synonym for Green Design) primarily focuses on combining environmental and economic advantages through good design solutions. It uses a systematic approach that aims to integrate environmental aspects

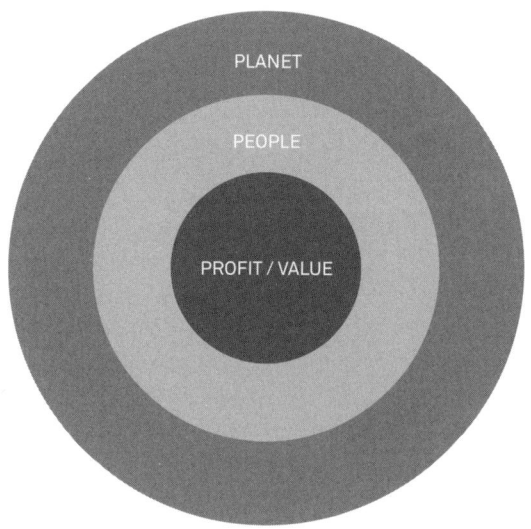

2 Sustainable Development 2.0: without natural environment (Planet) neither society (People) nor economy (Profit/Value) are possible (Tischner et al. 2015).

into the product planning, development, and design process as early as possible. Therefore, along with classical product development criteria, such as cost effectiveness, safety, reliability, ergonomics, technical feasibility, and of course aesthetics, environmental aspects become requirements along the whole life cycle and system. The Ecodesign designation states that with the help of a sound design approach, ecology and economy can finally be united.

When compared with Ecodesign, Design for Sustainability generally concerns itself with larger systems, e.g. production and consumption systems, and more radical questions.

Social, humanitarian, or fair design focuses above all on social and ethical aspects and tries to integrate them as much as possible into a comprehensive design. Considered are factors such as the impact of goods in society, working conditions in the production chain, fair trade issues, poverty, and social injustice. Potential human rights violations, health issues, gender equality and other social aspects are analysed and possible design solutions are sought that can positively impact the social environment. The United Nations Millennium Development Goals (MDGs) and the updated Sustainable Development Goals (SDGs) provide a good orientation for key areas of global activity.[6]

As depicted in Fig. 3, all other forms of environment-friendly or social design can be classified under the heading of Design for Sustainability aiming at creation of sustainable production-consumption systems.

That means compared with the mainstream understanding and practice of design today, we need to transition towards a much more radical approach that

Design for Sustainability
Sustainable Production-Consumption-Systems

Social Design

Social Innovation

Humanitarian Design

Fair Design

Base of the Pyramid Design:
projects with under-represented groups

Ecodesign:

- Use of sustainable materials and design for materials efficiency

- Optimise energy use and design for energy efficiency

- Design for zero toxicity and risk

- Design for optimal lifetime

- Design for circular economy and zero-waste

- Design for efficient transportation and packaging

questions the usefulness of anything that we design for as many people as possible and the related impact on the natural environment around us. This is a rather utilitarian approach or to say it with Spock from *Star Trek:* 'The needs of the many outweigh the needs of the few; or the one.' One could even go one step further and adapt Kant's Categorical Imperative: 'Act only according to that maxim whereby you can, at the same time, will that it should become a universal law' (Kant 1785), and suggest: 'Design only those kinds of goods (products and services) whereby you could, at the same time, want that they become universal goods that are used by all people around the world.' And then answer questions about sufficient availability of resources, energy, and land, and other consequential impacts on the natural and social environment.

Of course design also shall make people happy and make life more aesthetically pleasing etc. thus it has hedonistic elements. But it cannot be accepted any more that often designers make a few people happy with their designs but at the cost of severely damaging the natural environment and also lowering the quality of life for many many more people.

This new design approach not only needs to be more radical, but also could be much more democratic and participatory which leads us to open and crowd-based innovation and design, discussed in the next paragraphs.

Sustainable Goods (Products and Services) Meet the Following Criteria:

- Useful: fulfil a meaningful social function, solve a problem
- Efficient and effective: in energy, land, and resource use
- Solar: use renewable energy – solar, hydro, wind, geothermal, muscle strength, or sustainably produced biofuels
- Safe: are risk-free and 'fool-proof', ergonomic and pollution-free, harmless to humans and nature
- Appropriately durable: depending on the function (short or long-lived) lifetime should always be appropriate, when short-lived they should be particularly cyclical (see below)
- Cyclical: waste can be used as a nutrient in the same or different technical or natural cycles
- As regional as makes sense: low transport and packaging efforts (but have in mind that some goods can be produced more efficiently in other regions and it can make sense to support local economies in other countries)
- Social: good for the socio-cultural environment, enhance quality of life, safeguard jobs, comply at least with International Labour Organization (ILO) standards and are produced with (regionally) acceptable working conditions
- Valuable: reasonable value for money, appreciated by the user, help ensure the long-term economic viability of the provider

All these features must be considered for the entire life cycle and system of goods. It can be difficult to meet all the criteria in the design and development process equally well, e.g. regionalisation vs. efficiency – often compromises must be made to get the most feasible and marketable combination of environmental, economic, and social benefits.

Open Innovation, Open Design and the 'Crowd'

Open Innovation is a term originally coined by Henry Chesbrough meaning

the use of purposive inflows and outflows of knowledge to accelerate internal innovation, and expand the markets for external use of innovation, respectively. [This paradigm] assumes that firms can and should use external ideas as well as internal ideas, and internal and external paths to market, as they look to advance their technology (Chesbrough, West, Vanhaverbeke 2006).

One shortfall of Open Innovation activities so far is that they are still very much directed towards one company and its benefits but not to the benefit of society as a whole. A different and much more interesting understanding of Open Innovation related to Sustainability is to seek innovation activities that involve diverse stakeholders, actors, and experts and start with a problem and a societally relevant need or demand. For instance, besides companies, innovation projects can also be initiated by NGOs, municipalities, communities, citizens, networks etc. – even one person, and the problems tackled could deal with serious sustainability issues.

Many online platforms have been emerging recently that open up creative and economic activities to the so-called 'Crowd' using the following methods and tools, many of which offer new opportunities for designers as well to bypass the common dependence on a commercial client in their design projects and for their income:

- Crowdsourcing, is a term composed by the words 'crowd' and 'outsourcing' that indicates the act of taking tasks usually performed by contractors (or employees) and outsourcing them to a specific community of people (the crowd) in systems of mass production (Howe 2006). This means in simple words that the general public (the crowd) is invited and enabled to generate contributions (ideas, innovations, concepts, services, products etc.) for a company or other organisation or the crowd itself.
- Crowd-voting, the public or a specific community (the crowd) votes for challenges, solutions, products, designs to identify which of a number of solutions shall be selected. Crowd-voting is also a new form of target group research used by companies to get insight into public opinion.
- Peer-to-Peer Production (p2p), is a new form of production (of goods, contents, or services) that involves members of communities in an organised way. It's a 'coordinated, (chiefly) internet-based effort whereby volunteers contribute project components, and there exists some process to combine them to produce a unified intellectual work' (Benkler 2006).
- Crowdfunding, an approach for generating funds by asking the general public or a specific community (Crowd) to invest in or sponsor activities, such as implementation of new solutions or products, via online platforms, for instance used by musicians and artists to fund their productions. Different forms of crowdfunding are distinguished such as, donation-based or reward-based crowdfunding (crowd-donation, pre-sales), crowd-lending, crowd-investment.
- New Business Models, emerging via online platforms, e.g. Sharing Economy models, grassroots economies that move the focus from mass production to individualisation and production on demand, and to ethical, personal, political, and sustainable values of goods, mass customisation, co-design and socio-preneurism (social and sustainable entrepreneurism) (Gassmann 2010).

These new emerging methods and tools for Open Innovation and crowdsourcing, offer a growing group of well-educated creative professionals opportunities to be engaged in their own creative projects or team up in temporary project groups to carry out divers projects while not being employed or paid in traditional ways by companies. In addition, the Maker Movement and FabLab activities (see makezine.com) or craft oriented do-it-yourself sites (like www.craftster.org or www.etsy.com) support these new bottom-up and participatory open design movements. This democratisation of design and production can lead to positive impacts in terms of Sustainability – but also very negative developments. For instance, comparable to the implementation of desktop printers for everybody that increased the paper consumption drastically, the widespread implementation of 3D printers and easy to use 3D design software could lead to similar negative effects. One role of designers in this context could be to support and train the amateur DIY-actors, Makers, Fab-Labbers and others to ensure high-quality designs and include sustainability aspects in their design and production.

One Case Study: The Sustainability Maker Project and its Platform www.innonatives.com

Recognising these new movements and the urgent need to develop more radical innovation approaches towards Sustainability, the European Sustainability Maker Project (www.sustainabilitymaker.org) has created the Open Innovation and Design for Sustainability platform www.innonatives.com, which specialises in innovation and design for sustainability. From the research carried out in the project it became clear that combining crowdsourcing, crowd-voting, crowdfunding and an Online Shop on a single platform to create radical innovation for sustainability has not been done before and is very promising. The online platform innonatives has then been developed by iterative agile software design processes, where modules and functionalities of the platform have been established, test used and then improved before the next parts have been added etc. Thus the platform is a living model of continuous improvement. The most important functions of innonatives were defined as

- carrying out innovation challenges (Open Innovation, crowd-sourcing and crowd-voting),
- a Crowdfunding module to generate budgets for implementation of sustainable solutions,
- an Online Shop to trade sustainable solutions,
- an Expert and Educational System, as well as helpful tools for sustainability evaluation and development of Intellectual Property Rights concepts, and

- a Solutions and Implementation Archive that offers case studies and information about all projects implemented by the platform.

One important aim of innonatives is to enable cross-cultural dialogue of a most diverse group of actors worldwide. The so-called 'crowd' collaborates for solutions development and exchange of knowledge. Many of the sustainability solutions needed today exist already somewhere around the world. Often actors in one part of the world simply do not know of the successful solutions in other parts. To learn about them can be very inspiring.

The challenges (innovation projects) on innonatives had to be designed in a simple and intuitive way: to post a challenge, a 'challenge owner' creates and describes his/her sustainability-related problem. By proposing this challenge, he or she asks the so-called Solvers (creative community of the platform) to develop solutions. Solvers post ideas and concepts individually or in collaboration. Crowd-voting and evaluation takes place via the platform. The users who suggest the most promising ideas and concepts are asked to develop their ideas further and develop a more refined solution. At the end of the challenge process, all solutions are evaluated by the challenge owner, the experts (internal and/or external experts of the platform) and the crowd (crowd-voting).

By the use of crowdfunding and the Online Shop, the winning solutions can be implemented via the platform, or outside the platform alternatively, whatever is best and most efficient. All innonatives modules can be used independently from each other: the innovation process, the crowdfunding module and the Online Shop. However, the platform will ensure the quality and focus of all projects initiated as well as that all outcomes include a high level of sustainability.

The Sustainability Maker project team chose to have two challenge modes on the innonatives platform: open or closed innovation challenges, with two types of Intellectual Property Rights (IPR) approaches. The open challenges (open to the public) adopt the Creative Commons licence 'Attribution-NonCommercial-ShareAlike 3.0 Unported (CC BY-NC-SA 3.0)' which is the default open licence of innonatives that users accept upon registration. This Creative Commons licence means that users broadcast and share their solutions publicly; every innonatives visitor can see them and also use and develop them further, but they always have to refer to the originator of the solutions, who owns the IP rights. If an actor wants to use any open solution commercially, a (financial) agreement has to be made with the original inventor of the solution.

In closed challenges the aim is to make it possible to protect IPRs e.g. for companies that like to commercialise the outcomes. Closed challenges receive a password-protected closed challenge innovation space and specific rules for Copyrights and Intellectual Property Rights distribution have to be created and published by the challenge owner and accepted and signed by users before entering into this closed challenge innovation space. This licensing process is transparent on the innonatives website and accepting it is a precondition for participation. Thus all users know what their rights and duties are, when joining any kind of innonatives challenge. As Dickinger and Stangl rightly state: 'Trust is important for information websites, but even more essential for transaction sites' (Dickinger, Stangl 2013). Therefore, innonatives has developed an IPR contracting module that helps challenge proposers to define fair IPR models and contracts for the challenges.

Figure 5 shows the basic process and modules of the innonatives platform.

Sustainable innovations are defined on innonatives as new products, services, and production-consumption systems that fulfil customer needs and further drastically enhance the social and environmental performance along the whole life-cycle and service or production-consumption system. Sustainable Innovation includes not only technological but also social innovation, changing behaviour, and how citizens organise daily routines. Social innovation unlike technological innovation does not necessarily need expert knowledge but can be created by non-expert groups with a good knowledge of the actual situation and the needs and demands of the actors involved. This has been proven during the testing of the innonatives platform.

Two critical factors for using the innonatives platform were also defined in the test period as language barriers and the connection between online and offline communities: in order for participants to be able to communicate with each other, a shared language is critical. Currently, the majority of communication on the innonatives platform is in English. This has been an obstacle for some users. A temporary solution was the integration of an 'Online Translator' tool, but these are still of relatively low quality. In the long run, the platform shall be available in several languages. It is already possible to open a challenge in another language than English. However, participation is then partly limited to people speaking the language featured in the proposed challenge, excluding all others. Sometimes this effect can be desired by the challenge owner.

The connection of the international online community with offline stakeholders and participating groups can be established by using 'ambassadors' e.g. design professors and students that take an innovation project on for one term as has been done in the 'Kitchen Challenge' that took place in a favela in Curitiba, Brazil, where the Federal University of Paraná took over the role of ambassadors.

One helpful success factor when forming a good online community is the similarity of users, which 'refers to the extent to which potentially relational partners perceive or feel psychologically, morally or emotionally close or accustomed to each other' (Antikainen 2007). This also counts for similarities in experience and background. Being an Open Innovation Platform for sustainability issues, innonatives brings together users that have at least one thing in common: the wish to create sustainable solutions and a better world. They all work towards a similar goal. This is

5 The standard innonatives process (Tischner and Beste 2016)

SUSTAINABILITY PROBLEMS

CROWD-VOTING

SUSTAINABILITY CHALLENGE

CROWDSOURCING

SUSTAINABLE SOLUTIONS

CROWD-VOTING

EXPERT PANEL

BEST SUSTAINABLE SOLUTIONS

CROWDFUNDING　　MARKETPLACE　　AUCTION

IMPLEMENTATION

- The **Sustainable Communication with Fun** Challenge was asking for an interesting and inspiring video clip to explain how sustainability innovation and design, and innonatives work. Cash prizes were awarded and the winning video went online.

- The **Kitchen** Challenge was asking the crowd to design a product produced from reused wood pallets to improve the socialising in kitchens of low-income houses, for instance in the Aguas Claras community in Curitiba, Brazil. Challenge owners were University of Paraná, EcoDesign (an SME that offers furniture made from reused wood pallets), and Soliforte, a company dealing with recycling. Three winners were selected (first, second, and third prizes) and the best project has been produced, donated to the participating low-income community, and is now sold via the company EcoDesign. The designers will receive 50 per cent of net profit from the sold items.

- The **Sustainable Design with Coconut Soil** Challenge, was started by the SME Proflora and aimed at substituting peat moss use in European horticulture and gardening through waste materials from the coconut industry. Three winners were selected (first, second, and third prizes) and the company chose the second winner as the most interesting project for implementation. Currently the designer of this solution together with the company is working on implementing the solution, which will be available via the innonatives online shop, the online shop of the company and several retailers in Europe.

- The **Renewable Energy Based RADIOvelò** Challenge was dedicated to design a RADIOvelò as self-sufficient movable unit powered by solar energy, to be used as travelling-radio for events, aimed at promoting youth empowerment. One winner was identified, who then together with the challenge owners Politecnico di Milano and the Share Radio organisation started a crowdfunding project on innonatives to finance implementation.

- The **Exhibit a Sustainable Exhibition** Challenge by Politecnico di Milano was dedicated to design an environmentally sustainable system to exhibit students' projects in (design) universities, since design schools worldwide are often lacking environmentally sustainable approaches to student exhibitions. One solution has been identified as winner and will be implemented by the Politecnico's exhibition department.

- The **Water for Life** Challenge was a closed challenge by a large South American company on water-saving innovations. The large company and University of Paraná were involved in it and the challenge was asking the employees of the company to develop ideas and solutions on the subject. The project has been highly successful with a great amount of employee participation and engagement, so that the company decided to implement the best solutions and to repeat this kind of closed internal challenge more frequently.

- The **Radical Sustainability through Biomimetics and 3D Printing** Challenge was started by Bionic-Network Hessen, VDID (Verband Deutscher Industrie Designer e.V.) and econcept, and was searching for the most inspiring and visionary ideas on how biomimetics or biomimicry can be combined with 3D printing to generate more sustainable solutions. The nine best ideas received first, second, and third prizes, and were presented at a conference.

- The **Mute Bin** Challenge, another closed challenge by the large company ESE, Europe's leading manufacturer of temporary storage solutions for waste, was asking for radically new ideas to reduce noise pollution in waste collection in Europe by redesigning the typical waste bins. The company reserved the rights to patent and copyright the ideas, which they select for implementation. This will result in a contract with the designer of the selected solution.

6 Examples of Sustainability Challenges carried out on the innonatives platform

one of the most favourable success factors of the platform. It makes it unique, uniting all users in the same important mission.

Thus starting with a meaningful problem and finding creative solutions to radically improve the situation by involving a large group of stakeholders and then putting a team of actors together for implementation of the solutions is a very promising approach to faster development and implementation of truly transformational solutions funded by the people who benefit from them and a larger community that sympathises with them.

The Role of Research in Design for Sustainability

Researching during the design process, for instance about issues such as environmental, social, and economic impacts of products, services, and systems, about better and new technologies and materials, about new business and organisational models for new ways to organise daily life and work routines, about consumer habits, values, preferences, and different target groups in much deeper ways than 'creating personas' (as is frequently done by designers and design students) etc. are essentially important in Design for Sustainability. There are many helpful tools available to support these research and evaluation activities during the design process as is shown in Fig. 7.

Design for Sustainability and Open Innovation itself can become research activities, as they target larger consumption production systems and search for radically different solutions and disruptive innovations. For instance, the innonatives challenge on combining 3D printing with biomimetics (biomimicry) to create radical sustainability solutions was looking for these disruptive innovations and involved much research in the process – the process was a research activity in itself.

And finally, there is much **research about** new and more radical methods to create sustainable, open technological, or social innovation needed as this is still a very young discipline. The innonatives Kitchen Challenge, for instance, was a participatory exercise involving community members of the low-income community in Brazil actively in the innovation process. This was facilitated by design students and faculty of the local University of Paraná and has been thoroughly researched by a PhD student of that university with highly interesting outcomes (Dickie 2016).

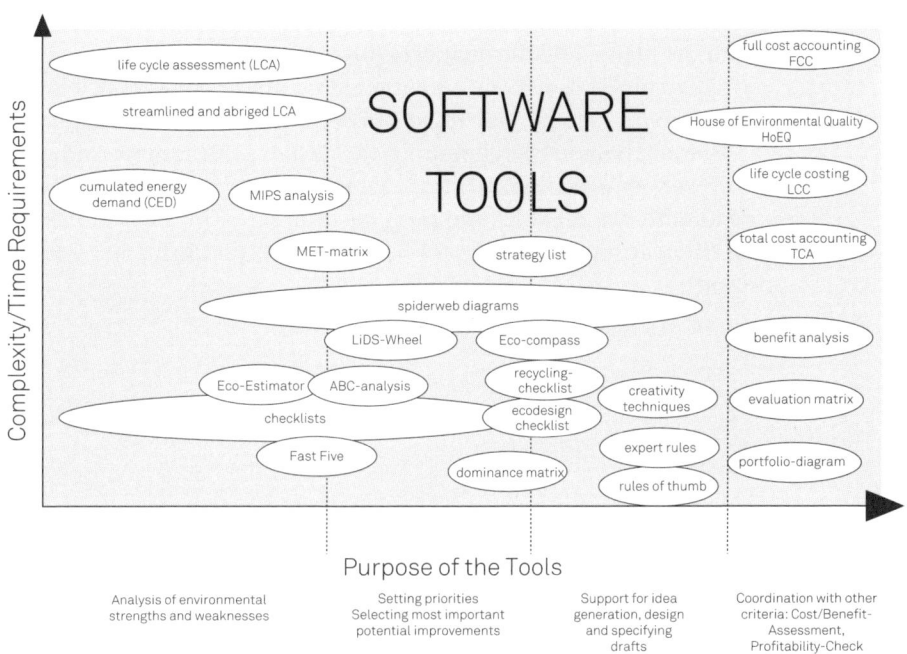

Figure with axes: vertical axis "Complexity/Time Requirements", horizontal axis "Purpose of the Tools". Title "SOFTWARE TOOLS".

Tools: life cycle assessment (LCA); streamlined and abriged LCA; cumulated energy demand (CED); MIPS analysis; MET-matrix; strategy list; spiderweb diagrams; LiDS-Wheel; Eco-compass; Eco-Estimator; ABC-analysis; recycling-checklist; creativity techniques; checklists; ecodesign checklist; Fast Five; dominance matrix; expert rules; rules of thumb; full cost accounting FCC; House of Environmental Quality HoEQ; life cycle costing LCC; total cost accounting TCA; benefit analysis; evaluation matrix; portfolio-diagram.

Purpose of the Tools:
Analysis of environmental strengths and weaknesses — Setting priorities Selecting most important potential improvements — Support for idea generation, design and specifying drafts — Coordination with other criteria: Cost/Benefit-Assessment, Profitability-Check

7 Helpful (research) tools and their purpose in Design for Sustainability (Tischner et al. 2015)

Conclusions

To summarise: there are already many interesting opportunities in Design and Open Innovation for Sustainability available and many activities are emerging and already carried out that offer the potential of establishing partly or fully a parallel system alongside the mainstream and dominant economic system and thus also interesting opportunities for designers. Crowdfunding and crypto currency schemes combined with the blockchain are new opportunities to establish financing outside the regular financial systems run by mainstream financial institutions and investors. Often these radical new concepts lead to start-up businesses and success stories and then also impact the mainstream markets. But all of this is still far away from being a mainstream activity among practising designers or in design education.

To tackle the urgent problems that humanity is facing, it would be really helpful if designers became more aware of the urgency of the situation and the magnitude of the problems, don't neglect this or fall into despair, but realise that it is actually a really interesting challenge and opportunity for creatives to question in every design project and assignment, how the solutions could contribute to improving the state of the planet and the quality of life of as many people as possible living on it.

This is not only needed and urgent but, at the same time, it is a lot of fun and a very satisfying activity, and at the end of the day makes the designer happy as well.

It would be extremely helpful, if the design community worldwide could come up with concerted actions to create more awareness about these issues and to promote a more responsible (and probably more exciting) way of designing. Joining Open Innovation and Design for Sustainability platforms like innonatives could be a start.

1   See http://www.footprintnetwork.org/en/index.php/GFN/page/footprint_basics_overview/.
2   See http://www.overshootday.org.
3   See Stockholm Resilience Centre, http://www.stockholmresilience.org.
4   See http://www.stockholmresilience.org/download/18.8615c78125078c8d3380002197/ES-2009-3180.
    pdf.
5   See UN Department of Economic and Social Affairs, Commission on Sustainable Development Ninth
    Session 2001.
6   See https://sustainabledevelopment.un.org/?menu=1300.

References

Antikainen, M. J. (2007). *The Attraction of Company Online Communities. A Multiple Case Study* (Doctoral Dissertation). Retrieved 7 May 2015 from http://tampub.uta.fi/bitstream/handle/10024/67697/978-951-44-6850-6.pdf;sequence=1.

Benkler, Y. (2006). *The Wealth of Networks: How Social Production Transforms Markets and Freedom.* New Haven, CO: Yale University Press, 277.

Chesbrough, H. W., West, J., Vanhaverbeke, W. (2006). *Open Innovation: Researching a New Paradigm.* Oxford: Oxford University Press.

Dickie, I. B. (2016). *Crowd-Design for Sustainability: A Reference Model Proposition.* 11º Congresso Brasileiro de Pesquisa e Desenvolvimento em Design in Curitiba, Brazil.

Dickinger, A., Stangl, B. (2013). 'Website Performance and Behavioral Consequences: A Formative Measurement Approach'. *Journal of Business Research* 66 (6), 771–777. DOI: 10.1016/j.jbusres.2011.09.017.

econcept, Material Sense, Van Markoviec, TU Delft, Cartesius Institute (2009). Final Report, *Green Fashion Project.*

Gassmann, O. (ed.) (2010). *Crowdsourcing.* Munich: Hanser Verlag.

Gesetz zur Förderung der Kreislaufwirtschaft und Sicherung der umweltverträglichen Beseitigung von Abfällen (Kreislaufwirtschafts- und Abfallgesetz – KrW-/AbfG), 27 September 1994.

Howe, J. (2006). 'The Rise of Crowdsourcing'. *WIRED magazine* 14 (1–4).

Kant, I. (1993) [1785]. *Grounding for the Metaphysics of Morals.* Translated by Ellington, J.W. (3rd ed.). Indianapolis, IN: Hackett, 30.

Packard, V. (1960). *The Waste Makers.* New York: David McKay.

Papanek, V. (1985). *Design for the Real World: Human Ecology and Social Change.* Revised edition. Chicago, IL: Academy Chicago Pub Ltd.

Papanek, V. (1995): *The Green Imperative: Ecology and Ethics in Design and Architecture.* London: Thames & Hudson Ltd.

Ressourcenkommission am Umweltbundesamt (KRU) (ed.) (June 2014). *Ressourcenleicht leben und wirtschaften. Standortbestimmung der Ressourcenkommission am Umweltbundesamt* (KRU), Berlin/Dessau.

Stebbing, P., Tischner, U. (2015). *Changing Paradigms*. Download for free http://www.cumulusassociation.org/changing-paradigms-designing-for-a-sustainable-future/.

Tischner, U. et al. (2015). 2nd edition. *How to do Ecodesign*. Available as e-book and iBook; e-book: https://www.umweltbundesamt.de/publikationen/how-to-do-ecodesign; iBook: https://itunes.apple.com/de/book/how-to-do-ecodesign/id1135764180?mt=11.

Tischner, U., Beste, L. (2016). *State of the Art of Open Innovation and Design for Sustainability*. Singapore: Springer.

UN (United Nations) (1992) Agenda 21: *The Earth Summit Strategy to Save Our Planet*. Document E.92-38352. New York: UN.

UN Department of Economic and Social Affairs, Commission on Sustainable Development Ninth Session, 16–27 April 2001, New York: *Indicators of Sustainable Development: Framework and Methodologies, Background Paper No. 3*. Prepared by: Division for Sustainable Development, http://www.un.org/esa/sustdev/csd/csd9_indi_bp3.pdf.

UNEP (2010). *How Close Are We to the Two Degree Limit?* Information Note.

World Commission on Environment and Development (1987). *Our Common Future*. Oxford: Oxford University Press, 27.

CROSSING THE BOUNDARIES OF PARTICI-PATION, ACTIVISM, PARADIGM CHANGE, AND INCUBATION: ON THE EDGE OF DESIGN FOR SOCIAL INNOVATION AND SUSTAINABILITY

Anna Meroni

The discourse on the connection between design and social innovation is today rich and animated: a decade since it began (Thackara 2005; Manzini and Meroni 2007; Davies and Simon 2013a; Manzini 2015) we can detect signals of maturity that are also widely reflected in several research programmes recently funded by the European Commission (EC 2011 and 2013b).

This chapter presents a conceptual framework, drawn from diverse applied research projects in design for social innovation, analysed through the lenses of some of the most debated theoretical approaches around design, innovation, and social issues in the broadest sense.

The research projects considered here come from the experience of the POLIMI DESIS Lab, the Politecnico di Milano based design team belonging to the international DESIS Network (Design for Social Innovation and Sustainability). This is a cultural association comprising several design universities and schools dealing with this topic, which has accumulated remarkable knowledge since its foundation in 2008, making it a privileged standpoint from which to reflect on the intersection of design and social innovation.

Design for Social Innovation: Evolving Knowledge

There is a growing convergence towards a shared understanding of what social innovations are: they are increasingly recognised as solutions that are social in their ends and in their means, based on new social forms and economic models, and are both good for society and enhance its capacity to act (Manzini 2015; Gabriel 2014; Davies and Simon 2013a; The Young Foundation 2012).

Social Innovation is a multifaceted phenomenon that can be read through different lenses (Nicholls, Simon, and Gabriel 2015): with regards to the discipline and the approach of Design, some key concepts can help create a conceptual framework for the ongoing research and practice in design for social innovation.

These key concepts exemplify diverse design approaches to innovation and to social issues in a broader sense that, as we will argue hereafter, can be gathered under the theoretical umbrella of 'design for social innovation'. Service design and strategic design are the disciplinary basis of the proposed conceptual framework: it is, in fact, acknowledged that many social innovations are forms of service innovation (EC 2013b), because they define new, regulated, forms of co-production of social benefits through solutions that imply the application of knowledge and skills by different parties (Meroni and Sangiorgi 2011).

Social innovations are, thus, behaviours and interactions that materialise value propositions aiming to generate social value. As forms of service, the methods and tools of service design can be applied to social innovation with different purposes: strengthening the capacity of innovators to work more effectively, be impactful, sustainable, and meaningful; assisting them to start up a venture; and fostering the capacity of society to be innovative and open to innovation. In other words, design can be a tool for empowering people and enabling them to make things happen (Selloni 2017; Manzini 2015).

While the theoretical discourse around social innovation is evolving, there is growing demand from international institutions such as the European Commission for a better co-production of knowledge between scholars and stakeholders from civil society, business, and government (WILCO 2013). This can be interpreted as a request for the creation of more integrated 'social innovation ecosystems', that's to say, locally rooted combinations of conditions, stakeholders, relations, and resources that collaborate to achieve a shared purpose able to generate public value. By 'public value' we mean 'the total societal value that cannot be monopolised by individuals, but is shared by all actors in society and is the outcome of all resource allocation decisions' (EC 2013a, 2). Despite the diversity of motivations and agendas, a social innovation ecosystem includes individual interests but has as its overarching goal the interests of the community. It is, therefore, a multi-stakeholder system comprising citizens, entrepreneurs, policy makers, and also researchers who can run experiments and projects in specific sectors while testing models and structures of collaboration (Meroni and Selloni 2018).

The way these stakeholders interact within an ecosystem can be encouraged, enabled and facilitated through different design approaches that can be visualised in the following Conceptual Framework (Fig. 1).

When speaking about 'design for social innovation', therefore, we are referring to a combination of design approaches that need to be investigated in their complexity and variety. Yet another two concepts also need to be discussed here, because they represent interpretations of design that overlap with that of design for social innovation: they are 'design for resilience' and 'social design'.

Resilience is an integrative construct that provides an approach to understanding how people and communities achieve and sustain health and well-being (Zautra, Hall, and Murray 2010). There are two specific notions to be considered from a design perspective: recovery (how well people bounce back and recover fully from challenge) and sustainability (how well people sustain health and psychological

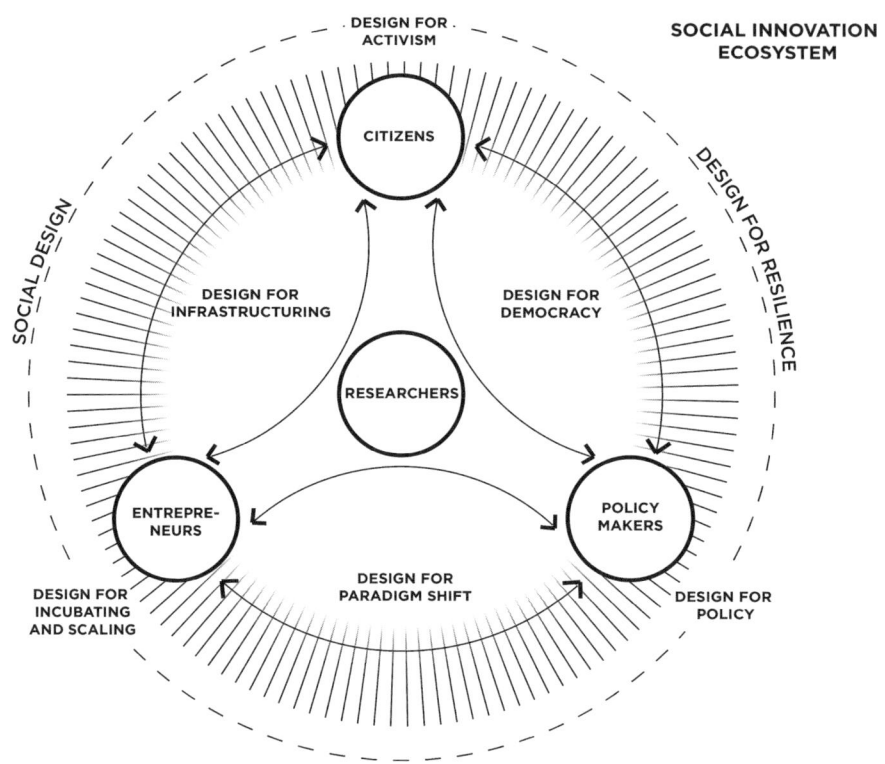

1 A Conceptual Framework of Design for Social Innovation (Credits: Meroni – Politecnico di Milano, Department of Design)

well-being in a dynamic and changing environment) (AA. VV. 2015). When referred to resilience, therefore, design is considered a way to empower people: overcoming difficulties by using creative thinking and a problem-solving mindset, looking at problems as opportunities and becoming open to change. Put in these words, we can see that 'design for resilience' overlaps very much with design for social innovation.

The other relevant concept is that of 'social design'. Adopting the definition of Victor Margolin (2012), it is the design that satisfies the human needs that the market doesn't take care of, because they concern populations who do not constitute a class of consumers in the market sense. With regard to this, Ezio Manzini points out (Manzini 2014) how this is different from design for social innovation precisely because the focus is on particularly problematic situations that are not dealt with by the market or by the state. Design for social innovation, instead, produces solutions

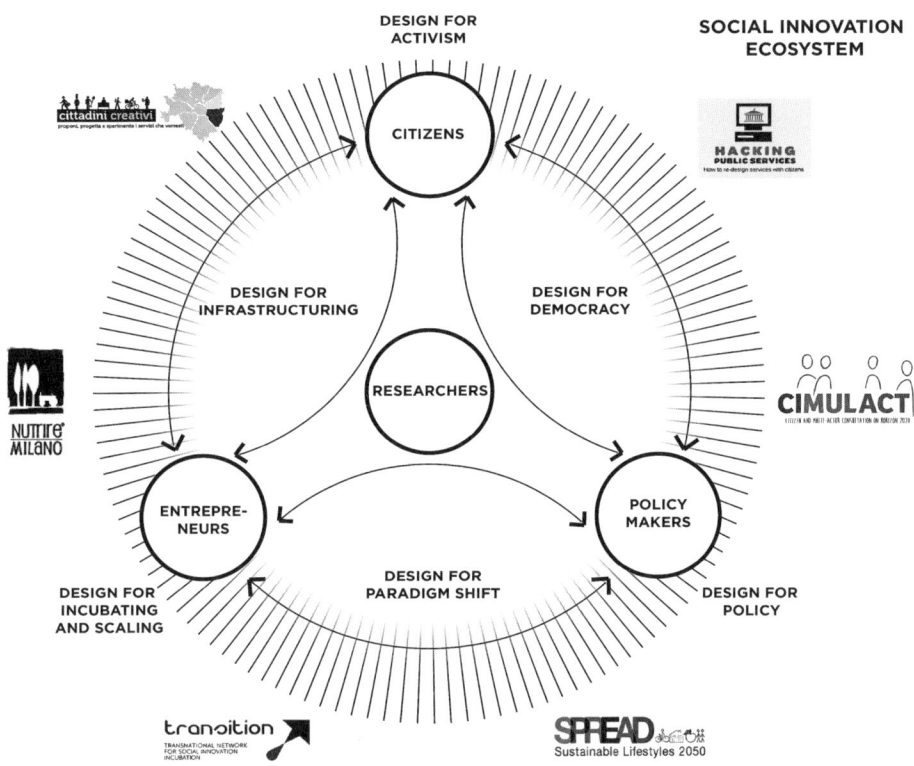

2 A Conceptual Framework of Design for Social Innovation, with research projects (Credits: Meroni – Politecnico di Milano, Department of Design)

based on new social forms and unprecedented economic models that are not necessarily thought up for very problematic conditions, but that attempt to reduce environmental impact, regenerate common goods, and reinforce the social fabric. In other words, they create public value.

Having clarified how design for social innovation meets other similar design approaches, let's see in detail how other, more specific ones, contribute to articulate and enrich its definition.

To do this, we can populate the Conceptual Framework of Design for Social Innovation with research projects that have been developed in the last few years by the POLIMI DESIS Lab: they will hereafter be discussed as experimentation that supports the argument of the framework (Fig. 2).

Design for Activism: Citizens

When the design intervention involves citizens, activism and participation are very pertinent concepts for social innovation.

On the one hand, so called 'citizen driven innovation' has been boosted in the last few years by ICT and the popularisation of smart devices:

Some cities are already applying open innovation models encouraging software developers to co-create tools and applications in collaboration with citizens and to tap into the knowledge generated in networks. [As a result], new models of citizen-driven innovation are emerging to re-define city services and how they are structured and organized, increasing the quality of public service delivery while also contributing to address the global challenges (Eskelinen et al. 2015).

In other words, technology has empowered ordinary citizens by offering them a way to make their voices heard (EC 2013a).

On the other hand, design is increasingly becoming a way to activate citizens (Davies and Simon 2013b; Scalin and Taute 2012): according to Markussen (2011), political design occurs when the object and processes of design activism are used to create 'spaces of contest'. Design activism is a disruptive interaction and aesthetic practice that, by introducing provocative artefacts into people's perception, invites active engagement, offers new ways of seeing and living, shows critical perspectives. The approach of design for activism, combined with a constructive intention and a pragmatic stance, can therefore configure an effective way for design to foster social innovation, particularly in the early stages. This is the case of the project Creative Citizens.

Creative Citizens

Creative Citizens (Cittadini Creativi) is an action-research project by the POLIMI DESIS Lab generated within the design doctoral programme of the Politecnico di Milano.[1] Located in a former farmhouse, Cascina Cuccagna, in the Zone 4 neighbourhood in the city of Milan, the programme was aimed at generating ideas with the citizens about services for the community. The former farmhouse, saved from abandonment by a group of citizens and local associations, is today a symbol of city activism and a permanent laboratory for civic participation (Selloni 2017).

Creative Citizens ran from February to June 2013 and, combining the tools of service design and scenario building, it enabled designers to work together with citizens to create a laboratory of solutions for everyday life, improving existing services and designing new ones (www.cittadinicreativi.it).

3 Design sessions from Creative Citizens – Cittadini Creativi (Credits: Cittadini Creativi Project, Selloni, Maggi – Politecnico di Milano, Department of Design)

The programme consisted of a series of co-design sessions around different need areas connected to daily issues and to services and places already existing in the neighbourhood. Each topic was developed through a sequence of three creative sessions:

- a *warm-up session*, to familiarise participants with the topic by presenting practices from various places in the world. It was aimed to inspire people, provoke debate, and offer visions of possible ways of life;
- a *generative session*, a collective brainstorming building on insights from the good practices and on the citizens' desires;
- a *prototyping session,* to move from the ideal service to a real one, identifying the assets that could be used.

The output of Creative Citizens was a first series of services co-designed with the people (*Augmented Time Bank, Object Library, Citizens' Desk, Facecook, Local Distribution System, Zona 4 Cicerons*) that can be regarded as public-interest services because they create a value that can be shared among the community and which does not benefit only few individuals, while responding to individual interests and needs. For example: less material waste, more social cohesion, higher citizen awareness, better personal competence, more care for the local environment, culture, and history.

After three years, these services are evolving in different ways (Selloni 2017), providing evidence of a successful process of citizen activation and empowerment towards social innovation (Cantù and Selloni 2013; Meroni and Selloni, 2018).

Design for Democracy: In between Citizens and Policy Makers

When considering how design works in the space between citizens and policy makers, we come to the connection of design with democracy.

Drawing from a reflection by Margolin (2012) about design and democracy, we can assume design 'for' democracy to be a design that addresses 'the opportunities for citizens to participate in democratic processes (focusing) on mechanisms and instruments for citizen engagement'. This requires two of 'the most important pillars of a democratic system, transparency, which enables citizens to be aware of the ongoing process of governance and the enforcement of laws, and participation, which is the opportunity to be involved in the process of government' (Margolin 2012) to be ensured by design, creating appropriate contexts and ways of interaction. Moreover, these principles are what characterise an open and collaborative government model, which implies better services for citizens more aware of their rights, with better access to information on public services and consequently higher expectations of service levels (EC 2013a).

Public services are services offered to the general public and/or in the public interest with the main purpose of developing public value [...]. Governments have to consider innovative new ways of developing and organising the public sector for creating public value (EC 2013a).

Here, social innovation plays a major part today, offering solutions that, thanks to ICT-enabled collaborative production, redefine the roles and relationships between professional, politician, practitioner, civil servant, expert, consumer, and citizen. The project Hacking Public Services exemplifies this approach.

Hacking Public Services

Hacking Public Services is a service design studio run in 2015 by a team from the POLIMI DESIS Lab,[2] within the Master of Product Service System Design run by the School of Design at the Politecnico di Milano. In partnership with the Municipality of Milan (more specifically the Council for Labour Policies, Economic Development, University and Research, and the Council for Social Policies and Welfare) a class of around forty-five international students was given the topic of designing public services addressing the above-mentioned transformations, using the potential of ICTs and exploring the collaborative power of design, production, and delivery of services.

The aim of the studio was to design a new generation of *open* public services in which citizens play a crucial role on the one hand, and on the other, government can support an ecosystem of actors generating public value (EC 2013a).

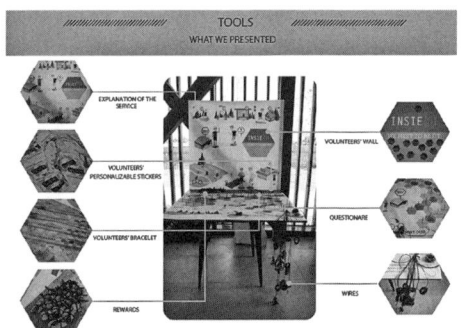
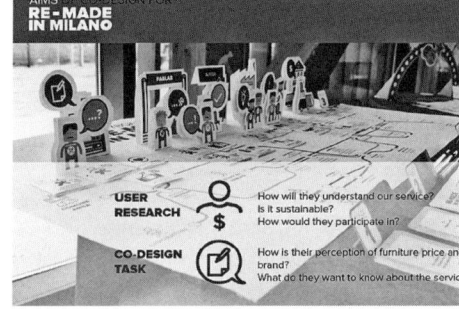

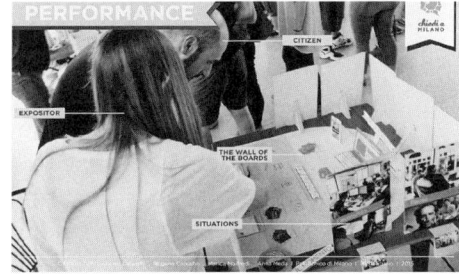

4 Co-design sessions from Hacking Public Services (Credits: Hacking Public Services Studio, Calvachi, Carvalho, Maifredi, Meda, Magnanimi, Massesi, Chen, Peskova, Bakir, Bonomi, Castellanos, Zhang, Basuki, Bolis, Cao, Hosseini, Mariani – Politecnico di Milano, School of Design, Master of Product Service System Design)

With the city of Milan as a case study, ten services for different areas of need (in-home assistance, new jobs, education, housing, new entrepreneurship, and sport) were designed with the citizens, exploring the potential of ICTs. A co-design session, organised with dedicated toolkits, was organised with the inhabitants of the Milanese Social Housing neighbourhood of Figino Borgo Sostenibile. The aim of the session was to stimulate the participants with initial concepts and provide them with the tools to enrich and develop them into more meaningful solutions for their everyday life.

The studio resulted in a presentation to the city councillor, creating an opportunity to open up a discussion about the future of the city's public services and lay the ground for further, more concrete actions.

Design for Policy: Policy Makers

When it comes to policy making, social innovation is being increasingly acknowledged as a significant driver of change. This reflects a paradigm shift in understanding the relationships between top-down and bottom-up, citizenship and governance, social and commercial entrepreneurship, profit and non-profit business (EC 2013b). As said, design has proven itself to be able to contribute to societal transformation by fostering activism from the bottom-up and by advising policy from the top-down (Meroni and Selloni 2018).

Moreover, through forms of 'cultural activism' (Fuad-Luke 2009) it has also proven to be an activator of community engagement and civic commitment around shared challenges (Bason 2014) and, according to the European Commission (DG Regional and Urban Policy and DG Employment, Social Affairs and Inclusion 2013b), open-innovation and open-government policies can contribute in creating a cultural humus (new ideas and expectations on the life) in society that can make people more receptive to innovative proposals.

Nevertheless, policy measures are needed to influence the regulatory framework in order to make it favourable to social innovation and allow an innovation to become a practice. Experimentation with civil servants and civic organisations is crucial to solve problems and to envision new possibilities for future systems (Bason 2014). Likewise, new research and innovation policies are needed to open the way to further social innovation. Design can contribute to building scenarios able to orient different kinds of policies, including those for research and innovation. It also contributes by materialising scenarios through solutions which suggest strategic socio-technical opportunities and by orchestrating participatory and multi-stakeholder processes to give voice to different social parties. This is the case of the international project CIMULACT, funded by the European Union.

CIMULACT

CIMULACT (which stands for: Citizen and Multi-Actor Consultation on HORIZON 2020) is a three-year project that ran from 2015 to 2018, funded by the programme Horizon 2020 of the European Union, operating to create visions and scenarios that connect societal needs with expected future advances in science and their impact on technology, society, and environment. It aimed at providing concrete input to the current research programme of the European Union (Horizon 2020), through recommendations and policy options for responsible research and innovation (www. cimulact.eu).

The core of the project lies in the engagement of citizens, along with a variety of other actors, in redefining the European research and innovation agenda and thereby make it relevant and accountable to society. The beginning of the research project saw more than 1,000 citizens in 30 countries in Europe formulate their visions for desirable sustainable futures: this was the basis for further debates and consultations with other stakeholders (experts, influencers, researchers, entrepreneurs, policy makers) to transform them into recommendations for future research and innovation policies and topics.

In CIMULACT, consultations were not necessarily conducted with a design-driven approach: the POLIMI DESIS Lab[3] was a core partner with the role of contributing to conceive the co-design actions where the visioning and 'thought-provoking' factor seems to be more crucial. In particular, it has been leading a large co-creation workshop with the aim of formulating research questions on the basis of needs extrapolated from the citizens' visions. In two days, more than 100 participants representing the diversity of stakeholders engaged in the project (from citizens to top scientists) have been guided towards the formulation of a first scenario for a research agenda articulated in twelve areas, corresponding to as many societal needs.

Adapting design-thinking tools to the specific circumstance, the flow of thoughts and the creativity of the participants have been guided from the interpretation of societal needs, to the creation of future research directions and the corresponding research questions. Participants, organised in twelve tables of around eight people, were prompted through an initial presentation in the form of a display of needs emerging from the citizens' visions. From this stage, a set of templates with a distinctive visual structure were used to support the thinking through complex logical steps, such as identifying key influencing factors for the future of a certain field, or ascertaining the present state of the art for both scientists and citizens.

An expert and trained coordinator per table helped to guide the groups and keep to the tight time schedule. The main contribution of design here was to conceive and make actionable the thinking flow of the workshop, so as to make it seamless and intelligible for all participants, regardless of their confidence with the situation, expertise, and education. In other words, the designer's job consisted in

5 The co-creation workshop of CIMULACT (Credits: CIMULACT Project – Politecnico di Milano, Department of Design, Lab Immagine)

creating the conditions for a truly *democratic* participation, making the language accessible, the process transparent, and allowing everybody to contribute by bringing her/his knowledge and opinion.

Finally, forty-eight research scenarios were generated, organised around twelve clusters of societal needs and with some initial visual input. These scenarios were then enriched through more face-to face and online consultations before being developed in the final research agenda.

Design for Paradigm Shift: In between Policy Makers and Entrepreneurs

There is growing interest in the 'transformative power' of social innovation. This is the process through which social innovation contributes to societal transformation that can be defined as the result of specific 'co-evolutionary' interactions between social innovations, system innovations, narrative of change (the discourses about innovation or change), and game-changers (macro-developments that change the rules, fields, and players of societal interaction). According to the research project TRANSIT (Wittmayer et al. 2015), social innovations are counter-narratives that propose alternative visions of the present and the future.

More in general, when talking about societal transformation, design is gaining a name as a catalyst for innovation in sectors where it has never before been viewed as such. 'Design is now being recognized as a significant factor contributing to the overall success of social innovation and sustainability projects', states Muratovski (2015). Moreover:

Contemporary problems associated with globalization, terrorism, epidemics, overpopulation, environmental issues, multiculturalism, and financial stability demand new solutions and unconventional approaches, and design is increasingly being seen as an agent of positive change (Muratovski 2015).

The capacity of design to envision radically new ways of doing is giving it a role in imaging alternative directions for the more sustainable futures of cities or regions and in supporting paradigm shifts of complex systems. Visions are powerful tools to engage stakeholders and citizens in longer transformation processes: service design helps to exemplify systemic changes at the level of everyday experiences and to materialise big shifts into tangible lifestyles and more intuitive business opportunities. Since design for social innovation is an attempt to foster transformations not only in people's behaviour but also in governance and in business systems, it largely benefits from methodologies that deal with collective processes of systemic learning and changes, such as those of scenario building.

Envisioning, visualising, and making future scenarios manifest through stories and services that represent how they will impact on daily life and what changes people will experience, is a way for design to work with communities that represent different interests and conditions. By doing this, designers move from a user-centred approach to a community-centred one, because communities of stakeholders become the interlocutor when acting for change on a systemic scale. By doing this, designers also reach into the organisation and participate in deeper transformation processes that lead to new business configurations and service models (Sangiorgi 2014). This is the case of the international project SPREAD, funded by the European Union.

SPREAD

SPREAD Sustainable Lifestyles 2050, a project funded under the Seventh Framework Programme of the European Commission, ran from 2011 to 2012 and was aimed at creating a narrative of change about the sustainable future of Europe, to inspire and orient transformations in lifestyles, productive systems, and policies.

The POLIMI DESIS Lab[4] was part of the research consortium. Its task was to organise the first envisioning process to start building scenarios of possible future sustainable living: scenarios seeking to explore the most extreme possibilities to help decision-makers plan for the currently 'unthinkable' (SPREAD 2012).

6 The issue cards of SPREAD (Credits: SPREAD Project – Politecnico di Milano, Department of Design)

From the perspective of design for social innovation, the methodology conceived and adopted in SPREAD is worth discussing. The scenarios were hypotheses of different futures, put in the form of stories about alternative environments and designed to highlight the risks and opportunities involved in strategic issues (Ogilvy

2002). They were created around current promising practices of social innovation. Therefore, a number of present-day practices selected from different fields were projected into the future by pushing some of their aspects to the extreme according to identified evolutionary trends. This generated a deck of more than fifty issue cards, organised in four areas, each of them depicting a small scenario of a future practice based on a contemporary one. The deck was used to feed a design workshop with the SPREAD researcher and several external experts who generated a set of visions of the future, translating sustainability into solutions for everyday life. This way, social innovation became a driver for imaging the future and setting up a conversation about it.

Design for Incubating and Scaling: Entrepreneurs

The ongoing debate about scaling and impacting social innovation is huge. According to the outputs of the European project TRANSITION,

the purpose of scaling is to grow social impact, rather than organisations. In this sense, scaling social innovation is different from scaling in the private sector, where it is more clearly about organisational growth (Gabriel, 2014).

In order to discuss incubation, first, we need to clarify the concept of scaling. Westley and Antadze (2013) have introduced a useful distinction:

- 'Scaling up', is the effort to connect the social innovation to opportunities in the broader economic, political, legal, or cultural context.
- 'Scaling out', is the effort to disseminate social innovation to benefit more communities.

Both are needed for a social innovation ecosystem to work well: the former is strictly related to the need to influence policies in order to allow paradigmatic changes to take place. The latter relates more to the idea of incubation as a process by which to increase the capacity of a venture to become self-sustainable and make an impact.

In the last few years design thinking has proved the validity of its contribution to the creation of methods and toolkits to incubate social innovation, yet we believe this is not the only way for design to operate (Cautela, Meroni, and Muratovski 2015). It also plays a key role in nurturing an innovation mindset in society at large. More specifically, we see design playing a double role in producing social innovation: a cultural role, with regard to desirability (meaning) and acceptability (society), and a technical role with regard to feasibility (technology), viability (economy) and ecology (environment). Another key contribution is then connected to the capacity of the design approach to connect or re-connect the enterprises with the community they target and refer to, in order to involve it in the co-creation and co-production process. This is particularly relevant when it comes to the social enterprise, whose connection and dependency on the society is key for the delivery of the services (Selloni and Corubolo 2017).

Cultural factors also tend to limit the technical possibility for a social innovation to scale: as Morelli points out (2014), scaling out by expansion is rarely effective, while scaling out by replication might work well. This implies scaling 'through "circles" or communities, each one guaranteeing trustworthy and meaningful interaction to its users, where the core service concept philosophy is preserved, but some variables are customized to the cultural and geographical traits of their communities' (Morelli, 2014). The exploration of how design can contribute to the incubation of social innovation was done through the international project TRANSITION, funded by the European Union.

TRANSITION

TRANSITION – Transnational Network for Social Innovation Incubation, a project funded under the Seventh Framework Programme of the European Commission, ran from 2013 to 2016 and was aimed at experimenting with both the technical and cultural incubation of social innovation, and creating a consistent, transferable body of knowledge by comparing the experiences of six Scaling Centres across Europe.

TRANSITION did not work only with conventional social enterprises or with start-ups aiming to become enterprises; it also worked with very informal groups of innovators who were not aspiring to build a formal business, but rather to become 'self-sustainable', i.e. to continue over time thanks to diverse 'adaptive' strategies for co-evolving within a changing environment and in response to changes in the interacting parts.

The POLIMI DESIS Lab,[5] partner in the project, set up a local scaling centre in Milan creating a partnership with local stakeholders, and played a major role in the conceptualisation of the common methodological path, the Social Innovation Journey, which was the main theoretical output of the project. Built on the experience accumulated with more than 300 social innovations during the project, it evolves the Nesta and Young Foundation's 'spiral' (Murray at al. 2010). In the TRANSITION approach, scaling means progressing at any stage of development of a social innovation, from being a simple 'prompt to act' to 'making a system change'. In fact, the Social Innovation Journey is a flexible incubation path that can be tailored to the needs of each specific innovation.

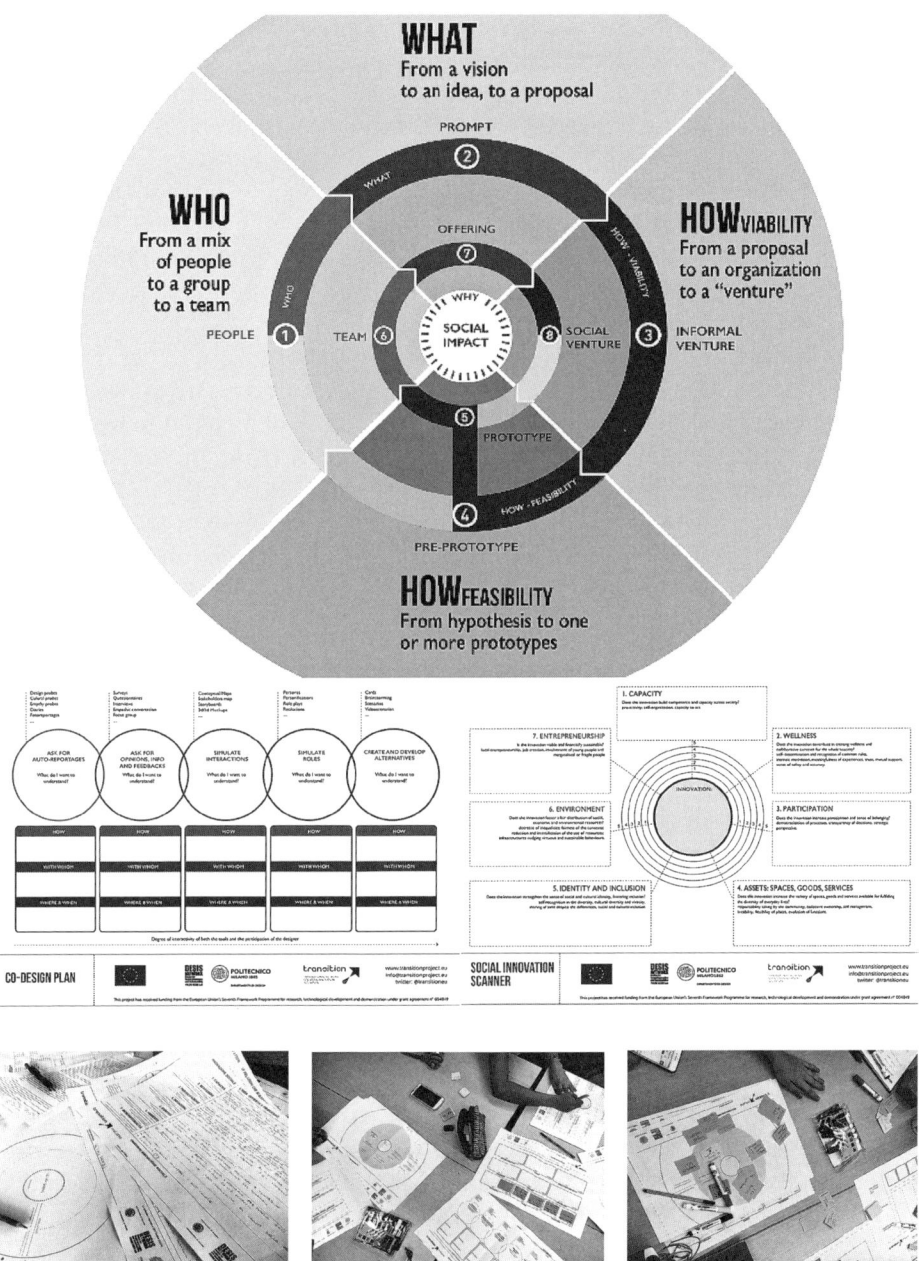

7 The Social Innovation Journey of TRANSITION (Credits: TRANSITION Project – Politecnico di Milano, Department of Design)

According to the experience accumulated in TRANSITION,

social innovations need people to work with, a proposition to test, customers, sources of advice, money, a place to work and mentorship (Miller & Stacey, 2014), but also a continuous alignment of intentions and visions with a broader and more fluid community, a motivational stimulus, the creation of a local culture and a methodological framework for operating and taking decisions. The SIJ is being built to integrate all these aspects in as simple a path as possible [...] organized so as to include methods and tools from enterprise incubation and acceleration and a wide set of service design tools (Meroni, Corubolo, and Bartolomeo, 2017).

The Social Innovation Journey (SIJ) consists of two circles of incubation and eight principal steps that touch on the main areas of incubation work corresponding to WHY, WHO, WHAT and HOW (TRANSITION 2016). Its general design can suit the scaling needs of any venture, but it is peculiar to social innovation because the tools and activities used in the different stages are thought to maximise the social impact of the activities, to facilitate a constant interaction with the social stakeholders through structured co-design activities, and to assess the social goal and the expected impact of the solution at any stage (Corubolo and Meroni, 2015).

Design for Infrastructuring: In between Entrepreneurs and Citizens

When considering the conceptual space between citizens and entrepreneurs, the discourse on design for social innovation has to do with enabling people to act in a purposeful and focused way. With this regard, the idea of 'enabling system' as a socio-technical system (an open-ended mix of technological/human/institutional elements) that allows social innovators to express their potential, act, and eventually make systemic changes (Jégou and Manzini 2008) is well known.

More recently, the term 'infrastructuring' has been used to refer to a continuous process of building relations with diverse actors and a flexible allotment of time and resources in order to foster innovation in society (Hillgren, Seravalli, and Emilson, 2011). Infrastructuring, as such, is a way to approach social innovation that differs from project-based design, because it is aimed at building relationships with stakeholders in order to enable them to act and create networks from which opportunities may arise. This approach gets closer to the idea of design for activism and integrates it with the technical dimension of incubation.

The project Feeding Milan (Nutrire Milano) is an exhaustive experimentation with the idea of enabling a large ecosystem of stakeholders to purposefully work around the specific topic of food.

8 The Ideas Sharing Stall of Feeding Milan – Nutrire Milano (Credits: Nutrire Milano Project – Politecnico di Milano, Department of Design)

Feeding Milan

Feeding Milan – Energies for Change (Nutrire Milano – Energie per il cambiamento) is an action research project that exemplifies this infrastructuring approach. Running from 2010 to 2013, it was funded by local institutions (Fondazione Cariplo, a bank foundation with the Comune di Milano and Provincia di Milano) and developed through a partnership between Slow Food Italia, the POLIMI DESIS Lab[6] of the Department of Design of Politecnico di Milano and the Università di Scienze Gastronomiche. Agriculturalists, gastronomists, and service designers worked together to create a network of interconnected services based on the principles of a short food chain, multifunctionality, and collaboration.

Connecting consumers in the town directly with farmers in the peri-urban area of the Agricultural Park South, on the southern edge of the town, the project has led to the creation of a network of services that work in synergy, according the principles of economies of scope and inspired by the idea of offering a rich and convivial experience (Illich 1973) to all the stakeholders (Meroni and Selloni 2018).

The project experimented with diverse initiatives that were all prototyped, but not all of them lasted over time. After ten years the most successful one is still the Earth Market: a farmers' market for local producers, organised according to the principles of Slow Food, where products are 'good, clean and fair'. As well as food sales, it includes educational workshops, taste laboratories, street kitchens and convivial tables where visitors can come together and eat.

One of the elements that have made this project relevant and instructive for design for social innovation was the systematic organisation of opportunities for design interaction between the stakeholders, including the citizens. This recalls the concept of infrastructuring, because the specific topics of the co-design activities were often instrumental and facilitated the creation of an enduring network of relations. With regard to this aspect the 'Ideas Sharing Stall' can be mentioned: it was a stand within the Earth Market where visitors and participants were invited by the design team to react and contribute to ideas for new services, giving comments and creative contributions. Over time an informal tendency has consolidated for the system of stakeholders, originated by the project Feeding Milan, to interact and 'design' initiatives for the future together.

Conclusions

By reflecting on applied research projects we have discussed how design for social innovation can be seen as a combination of a variety of similar design approaches.

Design is an acknowledged way to activate citizens: design activism is a practice that, by introducing provocative artefacts into people's perception, invites active engagement and offers new ways of seeing and living. If combined with a constructive intention, it is an effective way for design to encourage social innovation in the early stages. To become a practice, nevertheless, social innovation calls for changes in the regulatory framework, and therefore experimentation with civil servants and civic organisations is needed. With this regard, design can enable changes in policy by building scenarios and orchestrating participatory and multi-stakeholder processes to give voice to all of society and to engage stakeholders and citizens in longer transformation processes that can lead to big shifts in lifestyles and business models.

When speaking about the creation of new ventures in the field of social innovation, design has proven the validity of its contribution to incubation processes, acting on both the cultural and the technical factors that determine the success or the failure of an innovation. Finally, when incubation processes target society in general, besides specific projects, we can speak about 'infrastructuring' strategies. These are permanent processes of building relations with diverse actors in society in order to foster innovation and to enable them to act and create networks from which opportunities may arise. This is the way for design to create the conditions for social innovation to flourish and for society to be receptive to it.

The Conceptual Framework of Design for Social Innovation visually summarises the relations between the different approaches and actors, evidencing how researchers, and therefore also universities, can play the role of relationship facilitator and design interactions within a social innovation ecosystem.

1   The project was developed by the researcher Daniela Selloni.
2   The team comprised Anna Meroni, Daniela Selloni, Stefana Broadbent, Martina Rossi, and Serena Leonardi.
3   The team comprised Anna Meroni, Daniela Selloni, and Martina Rossi. Ana Maria Ospina Medina contributed to the design and organisation of the workshop.
4   The team comprised Anna Meroni and Marta Corubolo, with contributions from François Jégou (Strategic Design Scenarios), Francesca Piredda and Walter Mattana.
5   The team comprised Anna Meroni and Marta Corubolo, in collaboration with Matteo Bartolomeo and Matteo Boccia of Avanzi/Make a Cube, a Milan-based incubator.
6   The team comprised Anna Meroni, Giulia Simeone, Marta Corubolo, Daria Cantù, and Daniela Selloni, with numerous contributions from other researchers.

References:

AA. VV. (2015). 'Designed to Scale, Mass Participation to Build Resilient Neighbourhoods'. Available at: http://www.participatorycity.org/report-the-research/.
Bason, C. (2014). 'The Frontiers of Design for Policy'. In *Design for Policy*. Farnham: Gower Publishing Ltd., 225–235.
Cantù, D., Selloni, D. (2013). 'From Engaging to Empowering People: A Set of Co-Design Experiments with a Service Design Perspective'. *Social Frontiers: The Next Edge of Social Innovation Research*. Research papers for a major new international social research conference. Nesta.
Cautela, C., Meroni, A., Muratovski, G. (2015). 'Design for Incubating and Scaling Innovation'. In Collina, L., Galluzzo, L., Meroni, A. (eds). *Proceedings of CUMULUS Spring Conference 2015 – The Virtuous Circle: Design Culture and Experimentation,* Politecnico di Milano, 3–7 June, Milan: Mc Graw Hill. Digital edition available at: http://www.ateneonline.it/cumulusmilan/home.asp.
Corubolo, M., Meroni, A. (2015). 'A Journey into Social Innovation Incubation. The TRANSITION Project'. In Collina, L., Galluzzo, L., Meroni, A. (eds). *Proceedings of CUMULUS Spring Conference 2015 – The Virtuous Circle Design Culture and Experimentation,* Politecnico di Milano, 3–7 June, Milan: Mc Graw Hill. Digital edition available at: http://www.ateneonline.it/cumulusmilan/home.asp.
Davies, A., Simon, J. (2013a). *Growing Social Innovation: A Literature Review.* A deliverable of the FP7-project: TEPSIE. European Commission, DG Research, Brussels.
Davies, A. and Simon, J. (2013b). *The Value and Role of Citizen Engagement in Social Innovation.* A deliverable of the project: TEPSIE. European Commission, DG Research, Brussels.
European Commission – Directorate-General for Research & Innovation, Directorate 'Innovation Union and European Research Area' (2011). 'Research On Social Innovation. Inventory of Projects Funded under the EU Research Framework Programmes'. From: http://ec.europa.eu/research/social-sciences/pdf/project_synopses/ssh-projects-fp7-5-6-social-innovation_en.pdf.
European Commission – DG Communications Networks, Content and Technology (2013a). 'A Vision for Public Services'. From: http://ec.europa.eu/newsroom/dae/document.cfm?doc_id=3179.
European Commission – DG Regional and Urban Policy and DG Employment, Social Affairs and Inclusion (ed.) (2013b). *Guide to Social Innovation.* Working paper. From: http://ec.europa.eu/regional_policy/sources/docgener/presenta/social_innovation/social_innovation_2013.pdf.
Eskelinen, J., García Robles, A., Lindy, I., Marsh, J., Muente-Kunigami, A. (eds.) (2015). 'Citizen-Driven Innovation'. The World Bank and European Network of Living Labs / ENoLL. Available at: https://openknowledge.worldbank.org/bitstream/handle/10986/21984/Citizen_Driven_Innovation_Full.pdf?sequence=9.
Fuad-Luke, A. (2009). *Design Activism: Beautiful Strangeness for a Sustainable World.* London: Earthscan.
Gabriel, M. (2014). *Learning Methodology and Preliminary Framework.* A deliverable of the FP7-project: TRANSITION. European Commission, DG Research & Innovation, Brussels.
Jégou, F., Manzini, E. (2008). *Collaborative Services. Social Innovation and Design for Sustainability.* Milan: Edizioni Polidesign.
Hillgren, P. A., Seravalli, A., Emilson, A. (2011). 'Prototyping and Infrastructuring in Design for Social Innovation'. *CoDesign* 7 (3–4), 169–183.
Illich, I. (1973). *Tools for Conviviality.* New York: Harper & Row.
Manzini, E. (2015). *Design, When Everybody Designs.* Cambridge, MA: MIT Press.
Manzini, E. (2014). DESIS website editorial of 25 July 2014, www.desis-network.org.

Manzini, E., Meroni, A. (2007). Emerging User Demands for Sustainable Solutions, EMUDE, in Michel, R. (ed.) *Design Research Now: Essays and Selected Projects*. Basel: Birkhäuser, 157–185.

Margolin, V. (2012). 'Design and Democracy in a Troubled World'. Lecture at the School of Design, Carnegie Mellon University, 11 April 2012.

Markussen, T. (2011). 'The Disruptive Aesthetics of Design Activism: Enacting Design Between Art and Politics'. In Proceedings of the Nordic Design Research Conference 2011, Helsinki.

Meroni, A., Corubolo, M., Bartolomeo, M. (2017). 'The Social Innovation Journey. Emerging Challenges in Service Design for the Incubation of Social Innovation'. In Sangiorgi, D., Prendiville, A. (eds). *Designing for Service. Key Issues and New Directions*. London: Bloomsbury Publishing.

Meroni, A., Selloni, D. (2018). 'Design for Social Innovators'. In Walker, S., Cassidy, T., Evans, M., Twigger Holroyd, A., Jung, J., (eds.). *Design Roots*. London: Bloomsbury Publishing.

Meroni, A., Sangiorgi, D. (2011). *Design for Services*. Farnham: Gower Publishing Ltd.

Miller, P., Stacey, J. (2014). *Nesta … Good Incubation. The Craft of Supporting Early-Stage Social Ventures*. Available at: https://media.nesta.org.uk/documents/good_incubation_wv.pdf.

Morelli, N. (2014). 'Challenges in Designing and Scaling-up Community Services'. In Sangiorgi, D., Hands, D., Murphy, E. (eds.). *Proceedings of ServDes 2014. Service Futures*. Lancaster University, 9–11 April 2014. Linköping: Linköping University Electronic Press, 215–225.

Muratovski, G. (2015). 'Paradigm Shift: Report on the New Role of Design in Business and Society'. In *She-Ji: The Journal of Design, Economics, and Innovation* 1 (2). Available at: http://www.journals.elsevier.com/she-ji-the-journal-of-design-economics-and-innovation.

Murray, R., Caulier-Grice, J., Mulgan, G. (2010). *The Open Book of Social Innovation*. London: Young Foundation and Nesta.

Nicholls, A., Simon, J., Gabriel, M. (2015). *New Frontiers in Social Innovation Research*. London: Palgrave Macmillan. Available at: http://www.nesta.org.uk/blog/new-frontiers-social-innovation-research.

Ogilvy, J. (2002). *Creating Better Futures: Scenario Planning as a Tool for a Better Tomorrow*. New York: Oxford University Press.

Sangiorgi, D. (2014). 'Service Futures'. In Sangiorgi, D., Hands, D., Murphy, E., (eds.). *Proceedings of ServDes 2014. Service Futures*. Lancaster University, 9–11 April 2014. Linköping: Linköping University Electronic Press.

Scalin, N., Taute, M. (2012). *The Design Activist's Handbook. How to Change the World (or at Least Your Part of It) with Socially Conscious Design*. New York: HOW Books.

Selloni, D. (2017). *CoDesign For Public-Interest Services*. Research for Development series. Cham: Springer International Publishing.

Selloni, D., Corubolo, M. (2017). 'Design for Social Enterprises: How Design Thinking Can Support Social Innovation within Social Enterprises'. *The Design Journal* 20 (6), 775–794.

SPREAD 2050 (2012). European Lifestyles: The Future Issue. Available at: http://www.sustainable-lifestyles.eu/publications/publications.html.

Thackara, J. (2005). *In the Bubble: Designing in a Complex World*. Cambridge, MA: MIT Press.

The Young Foundation (2012). Social Innovation Overview. A deliverable of the project: 'The theoretical, empirical and policy foundations for building social innovation in Europe' (TEPSIE), European Commission – Seventh Framework Programme, Brussels: European Commission, DG Research.

TRANSITION Project (2016), TRANSITION SIJ Toolbox, Learning Outcomes. Brussels: TRANSITION. From: http://transitionproject.eu/learning-outcomes/.

Westley, F., Antadze, N. (2013). *When Scaling Out is Not Enough: Strategies for System Change*. Paper presented at Social Frontiers: The Next Edge of Social Innovation Research, 14–15 November, London Westphal. From: http://www.nesta.org.uk/event/social-frontiers.

WILCO (2013). Social Innovation Research in Horizon 2020. Position paper. June. Available at: http://www.wilcoproject.eu/wordpress/wp-content/uploads/WILCO-Position-Paper-SocInnov2.pdf.

Wittmayer, J. M., Backhaus, J., Avelino, F., Pel, B., Strasser, T., Kunze, I. (2014). 'Narratives of Change: How Social Innovation Initiatives Engage with their Transformative Ambitions'. TRANSIT Working Paper #4, October 2015. Available at: http://www.transitsocialinnovation.eu/content/original/Book%20covers/Local%20PDFs/181%20TRANSIT_WorkingPaper4_Narratives%20of%20Change_Wittmayer%20et%20al_October2015_2.pdf.

Zautra, A. J., Hall, J. S., Murray, K. E. (2010). 'Resilience: A New Definition of Health for People and Communities'. In Reich, J. W., Zautra, A. J., Hall, J. S. (eds.). *Handbook of Adult Resilience*. New York: Guilford Press, 3–29.

DESIGN AND INCLUSION: AN APPROACH TO ASPECTS OF INTEGRATIVE DESIGN

Tom Bieling

Now and again, important aspects of integrative design are detectable by virtue of design approaches that go beyond the cultures of the socioeconomic mainstream. This is true in particular since design contributes to the constitution of socio-cultural categories. As an example, let us take 'disability', a category that is reducible only with difficulty to specific bodily characteristics, but is affected to an equal degree by spatial, visual, virtual, or urbanistic, architectonic, in short: by design aspects. Approaches to the thematic field of disability, for example, nearly always proceed in the context of research in the medical and rehabilitative disciplines or from a sociological perspective. The focus of interest in this article, by contrast, is the role of technology and its design. The point of departure is the hypothesis that there exists a close interrelationship between technology and disability, as well as the assumption that technology demarcates the boundary zone between disability and normality, so that technology is involved simultaneously in processes through which disability is generated and interpreted. The aim of this essay is to demonstrate the ways in which societal and technological developments can potentially transform our perspective of disability. In the process, it should become clear that ways of defining issues that are purely technological in orientation are serious contributors to the 'problem'. The degree to which various benchmarks of design and social inclusion can be rendered more accessible to future discussions, both within and outside the design field, is demonstrated here with reference to four mutually complementary positions.

Background

To an increasing degree, demographic change on the one hand and technological progress on the other are leading towards a redefinition of disability and normality that contrasts with much past practice. In the context of growing life expectancies worldwide, first of all, we see an increase in 'the likelihood of acquiring a "disability" and/or being dependent upon care from others for extended periods of time' (Tervooren 2002, 1). Increasingly, as a consequence, disability is becoming a universal social experience. Questions concerning approaches to dealing with disability, of societal norms and values, must therefore be framed in new ways if we want to avoid displacing the majority of the population to the social margins (Hermes 2006, 28). On the other hand, technological auxiliary resources, i.e. prostheses and other assistance systems, are able to compensate for so-called physical weaknesses and deficits, which in turn results in new challenges when it comes to dealing with technologically driven bodily modifications. Illustrative of this is the current discourse around 'enhancements',[1] the 'cyborg' debate,[2] as well as independent philosophical tendencies such as 'transhumanism'.[3]

In light of potential and actual redefinitions of the human body, how is disability to be characterised? It seems advisable to avoid reducing the category of disability to bodily characteristics of the human, and to understand it instead as an aspect of a network of interconnected actions within which social processes, cultural constructs, and the designed environment are also of significance.[4]

Design and Disability

As disciplines that intervene in and shape our living environment, technology and design contribute in fundamental ways to the dissemination and stabilisation of the construct of normality. Whether in the form of ideals of beauty and stereotyped user definitions, which are scenarised and popularised by design and/or the media, or by means of product landscapes that are oriented towards the majority, and which are exclusionary in character despite this (or instead precisely because of it). As a consequence, there is a close interrelationship between disability and the design of technology (Bieling and Joost 2018a).

To date, we lack clear arguments for the existence of a relationship between design and disability. Inherent, meanwhile, to the concomitant critique of social relations, as well as of their causes and modalities, are claims concerning the necessity for their transformation. Design research is capable of responding meaningfully to these claims, in particular when design is understood to be based on the following aspects: first, design is conceived as being necessarily relevant to the larger society. Secondly, design is regarded as a discipline that possesses the potential to transform entities and circumstances (preferably positively), and which accordingly aspires to do so.

The Concept of Disability in the Context of Design and Technology

In connection with design, the difficulty of dealing conceptually with disability becomes explicit (cf. Bieling and Joost 2017). We encounter difficulty already with terms such as 'assistive technology' and 'assistive devices'. Evidently, the question of who must be helped or assisted already represents a problem. Those to whom such 'support' is addressed are unavoidably constituted as being 'in need'. At the same time, both the assistive resources and their designers appear as representatives of a principle of charity that is inherently hierarchical: those who require assistance inevitably stand in a relationship of dependency, and hence within a relationship of power, vis-à-vis those who offer assistance, or who are able to do so.[5]

A term frequently used in this context is the Anglo-American 'design for special needs', which has no precise German equivalent. Possibly for the same reasons that are adduced by critics of the English term, which they reject as patronising (cf. Pullin 2009, 2). Problematic here as well is the term 'medical engineering', which implies a pronounced focus on medical and technical aspects, thereby leaving aside social and cultural components entirely.

Universal and Inclusive Design

Particularly in relation to a rapprochement between the parameters of 'design' and 'disability', design theory and practice have developed a number of approaches, which circulate under various designations: to begin with, there is 'universal design' (Erlandson 2008; Herwig 2008), 'design for all', 'design for accessibility', 'barrier-free design', 'transgenerational design', and 'inclusive design' (Imrie and Hall 2001). According to Mitrasinovic, their ethical principles are roughly identical across regions – despite distinctions that are mainly terminological in nature (Mitrasinovic 2008, 419).

The concept of 'universal design' is closely associated with the US-American accessibility movement, and can hence be aligned chronologically with the European concept of 'design for all', whose origins can be found both in the 'democratic' design approaches of Scandinavian functionalism[6] (Klein-Luyten et al. 2009, 13), but can also be traced back to the ergonomic design of the 1960s (Kercher 2006; cited from Leidner 2007, 398).

That the basic aims of the above-named approaches are essentially in sync with one another becomes clear when we realise that at design academies, in political and commercial contexts, as well as in everyday language, these concepts are often used as synonyms,[7] listed in succession, or simply mixed together. Invariably, the core message seems to be: the resources of design are to be used in order to reduce difficulties for as many people as possible![8] An equally central credo can be traced back to the disabled architect and industrial designer Ron Mace, who introduced the term 'universal design' during the 1980s, accompanied by the following argument: design should consider all users. Not only the average user. And not only those who are designated as 'exceptional' or 'abnormal' (Mace 1985).

The Problematic Focus on Problem-Solving

Clearly, approaches such as 'universal design' are based on a socially oriented world-view. Nevertheless, the design outcomes associated with it may involve difficulties. This is true in particular when it is assumed that a design object that is comprehensible and usable by all is a) possible in the first place, and b) is necessarily the best of all of the available options.[9] In particular with technical devices, attempts to address a maximum number of user scenarios often lead to well-intentioned but badly made products, which are often 'multifunctional' in the sense of being overloaded with ancillary functions. Coming to light here is a logical error: the more variations of use and operation a product offers, the more potentially inclusive it must be. But if the degree of complexity associated with its use increases to the same degree, it seems doubtful that the inclusion argument can be maintained (cf. Pullin 2009, 67). Too much modularity and too many possibilities for adjustment also produce a higher degree of visual or functional complexity (for example), which can lead to misunderstandings or errors in use.

Here, theory and practice diverge, and at the same time demand and reality. The main question seems to be whether the aim is the realisation of a design of sub-areas that is *maximally* barrier free, or a design that is 100 per cent barrier free in all areas? The latter can probably never be implemented in a satisfactory way, i.e. with expenditures of resources and actual use value standing in an appropriate relationship. In fact, there exist products that are barrier-free to a high degree, but which are nonetheless far from fulfilling all of the criteria of universal design.

It is clear then, that the general interventionary potential which is inherent in design permits the elaboration of highly divergent results, which may be perceived – depending upon the target group addressed and the context of utilisation – as either positive or negative. This becomes especially clear with reference to the issue of stigmatisation that often exists in connection with the aesthetic appearance of medical aids. Many such projects are tarnished by the presence of a 'support stocking' aesthetic.[10] Often, such projects and products are conceived for emergency situations or hospital stays, with little consideration given in many cases to the actual lived realities of users. In too many cases, design for people with disabilities results from a one-dimensional conception of users, who are reduced entirely to their 'patient' status.

In a wider sense, this means that disability itself is regarded as a problem which must be solved. In many instances, a design approach that addresses the theme of disability explicitly operates within a clinical context and moreover from a 'medical-diagnostic' perspective, which not only fosters an 'excessively clinical' type of design that is perceived by everyday users as stigmatising, and hence often rejected, but also fuels the tendency of the dominant culture towards a problem-solving orientation.

The question however arises: to what degree is design for people with disabilities genuinely satisfactory as a problem-solving approach, and what new types of problems might it generate?[11] Accorded little attention to date is the circumstance that inherent to the complex network joining between 'design' and 'disability' is a design dilemma, one that represents an 'irresolvable problem' in Brock's sense.[12] This can be clarified with reference to a pair of brief theses.

Thesis 1: Dilemmas of Design

When it comes to the concept of disability, a purely traditional – i.e. problem-solving – design approach seems unviable. An approach in the spirit of a 'design *for* disabilities' excludes decisive aspects (or positions them explicitly in the foreground), thereby confronting designers with a dilemma: design *for* the disabled designs disability.

In concrete terms, this can be broken down in relation to at least two dilemmas: first, there is the question of the visibility versus the concealment of disability (guided, again, to begin with by formal-aesthetic concerns,[13] but is above all a question of the way in which disability is perceived by those affected as well as by others, and hence sociological in character). And secondly, a 'helping' approach can be criticised as paternalism.

With regard to disability, then, design finds itself trapped in a catch-22 situation. As soon as we attempt to design for people with disabilities, we necessarily find ourselves designing disability as well. Emerging in the process, as a rule, is more than one design option – about which, incidentally, the affected individuals may find it difficult to make decisions. Should a physical deficit be concealed, for example, or instead deliberately emphasised? Advantages and disadvantages are associated with each option.

For a long time, for instance, hearing aids were far too large, and hence visually conspicuous.[14] This could be perceived by wearers as stigmatising: one was perceived outwardly as hard of hearing, and potentially reduced as an individual to this impairment. Undeniable, on the other hand, is a (perhaps more positive) secondary effect: strangers were signalled immediately that they might want to speak louder than usual, or would at the very least understand why their interlocutor failed to respond immediately, and perhaps spoke in an unaccustomed manner as well.

The counter model in design terms consists in concealing the device, so that it is virtually undetectable or even invisible to others – not unlike contact lenses. This avoids having users reduced solely to their disability by others.

In the search for 'solutions' to the 'problem' of disability, consequently, design is faced with a conflict: on the one hand, there is the desire to be helpful; and on the other, design necessarily has a normative impact on the way in which social definitions and courses of action are manifested.

It amounts to a question, then, of the degree to which an ostensibly medically oriented design approach, i.e. one that attempts to compensate for a disability as discreetly as possible,[15] actually implies that disability is something that must be concealed or camouflaged (cf. Pullin 2009, 4). There is reason to believe that an approach oriented towards the social dimension of disability would generate different results. This seems all the more likely in light of the following, second thesis.

Thesis 2: Perspectives of Disability as a Driver of Innovation

A design approach not oriented exclusively towards compensating for disabilities can result in perspectives that go beyond applications that are restricted to the disability context, and hence become relevant to an expanded definition of target groups.[16]

Assumed in many cases for product development within the special needs sector[17] is a so-called 'trickle down' effect (cf. Pullin 2009, xiii). This circumscribes the familiar processes through which modes of function, productive methods, the processing of material, or other aspects of design go beyond 'mainstream design'[18] and are applied later in smaller markets (ibid.). At least as interesting, however, is the contrary effect which emerges when the thematic field around disability functions as a catalyst for new design approaches, hence opening up domains of action for more broadly diversified design cultures.

In the following, I refer to such context-spanning phenomena as cross-functional concepts. Meant here are design approaches for which disability is not the intended receiver, but instead functions as the point of departure for processes whose subsequent development may lead towards domains of application that are independent of this context. A special trait of such approaches would be a refusal to understand potential disabilities as deviations from the norm, as exotic phenomena or as deficits, and instead – on the contrary – regarding those characteristics that are associated with a specific disability as a 'normal condition'.

Trajectories of Cross-Functionality

As mentioned above, transfers into other contexts of use may proceed along a variety of trajectories. A comparatively direct form of transfer occurred with the typewriter, originally intended as a means of communication for the blind, and only disseminated subsequently among broader user groups. Ulrike Bergermann describes the success story of the typewriter as a prominent instance of the 'development of machines and technologies designed to overcome disabilities which at times become compatible with the needs of the majority of society' (Bergermann 2013, 19).

Another prominent instance of cross-functional concepts in the context of disability is the system of braille writing. Developed by Louis Braille in 1821, the original function of this 'writing for the blind' was to serve as a system of transmitting messages among Napoleon's troops; it allowed soldiers at the front to read communiqués in darkness without the use of illumination, which would have betrayed their positions to the enemy.

A third example can be found in the area of the audio book market, whose great success goes back to the development of 'books on tape', whose original concept involved making literary works accessible to blind people through the auditory mode. Today, audio books are still used by the unsighted. Meanwhile, the greater part of the substantial sales of audio books are generated by sighted customers, who take advantage of the option of absorbing books while jogging, driving, or going to sleep (and hence with closed eyes).

Disability as a Point of Departure and Target of Design

To date, the design discourse has accorded little attention to fundamental aspects of disability. This is all the more astonishing in light of the normative force exerted by design. Design decisions are often oriented towards principles of usability. The concept of 'usability', however, seems to be oriented towards a majoritarian principle: the aim of most usability tests is to find out what the majority of people say about a product, what they do with it, what they seem to think of it.[19] People with disabilities – to the extent they are perceived as 'deviating from the norm' – tend to be excluded from such majorities. Is 'usability' (to the extent that it genuinely targets majorities) hence a concept which automatically reinforces 'disability' – in the sense of a power relation – or makes it possible in the first place?[20] If one follows this argument, then 'usability' stands unavoidably in connection with questions of 'normality'. Particularly when it becomes clear that always inherent in usability processes, according to this argument, is a certain normative pressure, whose bases are for example statistical or functional in nature.

If it is assumed nonetheless that 'usability' is non-judgemental, one could pose the question, in a countermove, of how aspects of usability could be oriented more specifically towards human diversity (not only in terms of dis/ability, but also with regard to age, gender, cultural aspects, educational attainment, or social status).

This clarifies the degree to which design is able to approach the theme of disability from a variety of perspectives. It also becomes clear that manifold intentions can underlie both design stimuli and design processes; i.e. with 'disability' serving as either the point of departure or target of a design. Disability, then, can either function as an 'object' for which solutions must be developed, or as a point of departure for design approaches that need not necessarily be bound to the original context. Both interpretative trajectories can be integrated constructively with one another: concrete design approaches may be developed initially for a specific disability context, which in turn – through a new procedure, technology, or product – open up new fields of activity. While with the first variant, the designer's role is that of the problem-solver, the second variant makes possible a less restricted space of possibilities for design, but also an explorative domain of design research, with its delight in experimentation (Bieling 2019; Pullin 2009).

Design and Inclusion

Described and discussed now in detail will be the central role design plays – or can play – in relation to disability. While the title of this essay stresses *inclusion*, then this of course implies that the principle of inclusion cannot be reduced to the context of disability alone. Instead, it relates – like the term *diversity* – to all ways of life, social sectors, cultural difference, and so forth. The intention here is to open up such a broadened conception of inclusion from a design perspective and to open it up for discussion.

The goal, meanwhile, is to provide an improved breakdown of the diverse points of reference for design and inclusion for future discussions within and outside the design field. And at the same time, to disclose the fields of operation for designers that are deducible from these points of reference. Elaborated below with reference to four positions are the various foci and points of access that exist between design (research) practice on the one hand and inclusion on the other.[21]

The design *of* inclusion: here, it is a question of participation in the form of *institutionalised participation* (for example in relation to political parameters). Design *for* inclusion: here, it is a question of *participation through design outcomes* (which is to say, of design as empowerment, for example through the design of the tools for participation). Design *through* inclusion: here, it is a question of *participation in the design process* (for example in the form of participatory design or participatory research). And finally, design *as* inclusion: here, it is a question of *participating in design* in the sense of forms of participation that are inscribed in design (for example in relation to an altered visibility of marginal groups; but also in the form of (design) activism.

What then can design do, what are its areas of responsibility and spaces of possibility, when it comes to facilitating social inclusion, to practising it, allowing it, taking it up, promoting it? What tools can it provide that can be used to (better) channel the required processes – at least tentatively? The following responses to these questions provide insights into potential positions for integrative design.

Design *of, for, through,* and *as* Inclusion

For the sake of a future, discursive, and practice-relevant confrontation with design and inclusion, both within and outside the field of design research, I now propose four positions which concern the ways in which design and inclusion can be related to one another, how design can operate with an eye towards inclusion. This subdivision does not necessarily aim towards localising or naming new fields of activity for design, but should be regarded as an attempt to emphasise more explicitly the various crosslinks between design and inclusion in order to create a discursive basis for this thematic complex that is comprehensible and amenable to discussion both in the practical and theoretical fields of design as well as beyond.

The first position refers to a design *of* inclusion. This takes place in particular in and through the corresponding institutions and agencies which deal for example with the conception, implementation, consultation, or critique of the corresponding political, legislative, or infrastructural framework conditions (i.e. ministries, municipalities, the courts, the civil service, but also citizens' initiatives, NGOs, etc.). The influence of the executive institutions (i.e. at the municipal or ministerial levels) on the design of inclusion pertains for example to legislative and infrastructural aspects. The influence of the consultative institutions (for example, citizen initiatives) refers for instance to serving as a source of information for political-institutional procedures or providing advice and support to their executive organs in the sense of a lobby.

A design of inclusion and an iterative 'optimisation' of inclusive processes takes place directly within the institutions themselves or in collaboration with them. As strategic designers or service designers, for example, designers who participate in this way in the (re)design of institutions can contribute by improving the processes that are enacted there or will be in the future. Or as communication designers, they can work on making the communicability of these processes more transparent, comprehensible, or efficient, whether internally or externally. Here, inclusion can be fostered with the help of design knowledge and its methods of development and implementation, with design providing assistance to institutions that are charged or engaged with inclusion.

The second position of design with regard to inclusion is the field of design *for* inclusion. Here, design can function as a provider and shaper of tools, objects, information, platforms, and networks, systems, in short: things that are helpful and useful for purposes of inclusion, for example regarding applications relevant to everyday practices. Here, it is a question in particular of artefacts, products, or end-user devices which serve as 'tools for empowerment' for individuals and communities.

One example of this is the Lorm Hand (Bieling and Joost 2018b) that is described in the research project 'Lorm Hand – Communication Devices for Deaf-Blind People'[22]. Design for inclusion strives to provide and facilitate access, to dismantle barriers, to allow more people to take part in social processes, to provide

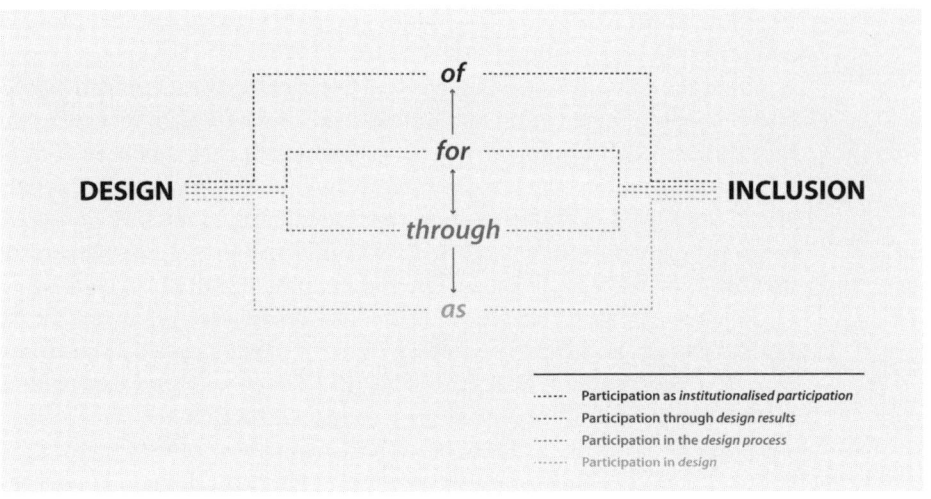

DESIGN ⋯⋯ *of* / *for* / *through* / *as* ⋯⋯ **INCLUSION**

⋯⋯ Participation as *institutionalised participation*
⋯⋯ Participation through *design results*
⋯⋯ Participation in the *design process*
⋯⋯ Participation in *design*

1 Points of reference for design and inclusion (Bieling 2019)

them with information and to make it easier to participate in decision-making processes, as well as initiating options for collaboration and networking (cf. Bieling, Martins, and Joost 2017). The potentialities of and developments in digital technology foretell an entire series of innovative forms of inclusion. Which does not mean that design for inclusion is restricted to the application and development of such digital technologies. In analogue domains as well, tools for inclusion can be effective throughout a wide spectrum of design disciplines – whether in the realm of product design, service design, through a focus on community building or civic infrastructures,[23] or through the provocative shaping of design as critical and speculative design (Dunne and Raby 2014).

Through the creation and provision of tools for participation, *design for inclusion* can also function as a driver of the above-described *design of inclusion*. And incidentally, also spur on the next position to be described, that of *design through inclusion*. This is especially the case where designers go beyond providing things, and – as demanded by Ezio Manzini – instead create framework conditions through which individuals, groups, and communities are able to generate their own solutions, working moreover in the absence of direct interventions by designers. Manzini refers to this as 'improving the space of possibilities', as the 'creation of an environment [as] enabling system' (Manzini 2017). A form of self-help, therefore, or a *design infrastructuring,* as Pelle Ehn characterises it (cf. Björgvinsson, Ehn, and Hillgren 2010). Such systems of empowerment ('enabling systems') can exist, for example, in the form of digital or analogue platforms, through the integration of social workers, and on the basis of hybrid forms of knowledge exchange, personal networks, and so forth.

The third position, design *through* inclusion, pertains to the processual aspect, that is to say the aspect of inclusion within the design process. Examples include in particular participative or co-design processes (cf. Ehn 2013; cf. Sanders 2002). But also innovative design perspectives like the cross-functional design approach formulated here can function as characteristic features of design through inclusion. Ideally, the principle of the non-exploitative participation of non-designers on an equal footing in the design process could be regarded as a precondition, or as an expression of a basic attitude that positions individual members of society on the same level, and which is based on a fundamentally democratic understanding of equality and equity.[24]

Where successful, in turn, *design through inclusion* is capable of informing and spurring on the other three positions: the *design of inclusion* as well as *design for inclusion*. And also the *design as inclusion*, described in the following.

The fourth position, *design as inclusion*, refers to a form of participation that is encoded in the design and thus manifested outwardly. This can relate, for example, to medial representations or – in the context of disability – to destigmatisation through a 'non-clinical' design approach. Through the creation and distribution of different iconographies, as well as through the application of different narratives, design can contribute on the one hand to increasing the visibility of minorities, without on the other hand singling them out as 'special', 'other', or 'abnormal'. Given the dilemmas facing design, as described above, this can mean a tightrope walk for designers. If the balancing act succeeds, designs can lend a voice to those who have been silenced – moving some distance towards 'normalising' them.

In this position, the forms of participation in design are also possible via different routes, for example in the form of (design) activism or protest, through which, with the help of design, concepts of diversity can be represented, and at the same time the interests of various minorities formulated, addressed, and communicated.

On these various levels, *design as inclusion* can contribute by posing critical questions concerning existing (power) relations, calling attention to possible alternatives, and hence instigating or moderating discussion. Here, inclusion is implemented on all levels and in all phases of the design process, including that of the results.

In comparison with the first three positions described above, the fourth position is perhaps at times more difficult to grasp, particularly since the examples assembled to illustrate it have been constituted in a deliberately open-ended and permeable way. *Design as inclusion* can perhaps be best understood as a principle that functions descriptively, and whose elaboration – like the other three positions – will hopefully be sharpened further in the coming years through interdisciplinary discussions.

Prospects

Self-evidently, the design of technologies and products always implicitly transmits role models and values. Design and the idealised images it disseminates, often unreflectively, are hence unavoidably political. Related to this is the question of which stereotypical image and product worlds are to be tolerated in a society. There is no doubt that the scope of action for designers also involves increasing awareness of the sociopolitical dimension of their design activities, in particular when it is a question of recognising the power of their designs to exclude, and of critically interrogating this power. Participation and inclusion are key principles that should play important roles as ethical norms in the domain of the design of technology. Since a design that is oriented uncritically and exclusively towards majoritarian principles sets up new hurdles for people with disabilities, a minimal demand should be to call into question entrenched models of understanding disability, for example, as deficiency. For this reason, participation and inclusion are indispensable aspects of integrated design, whose fundamental position should be situated within the realm of human-centred design and the questioning of systemic and putative realities. In the best sense, this position can be regarded as the point of departure for a constructive criticism and transformative processes.

1   The concept of *enhancement* pertains to the multifarious possibilities for optimising human – i.e. cognitive or physical – capacities, for example through medicines or implants. The ethical dimension of the increasing feasible and in fact practically implemented possibilities of reshaping human beings consists in particular in the wider implications for society and its conception of the human and the fact that more and more people feel the urge to perfect themselves (cf. Schöne-Seifert and Talbot 2009; Gesang 2007; Dickel 2011).

2   The term cyborg (derived from the English 'cybernetic organism') refers primarily to hybridisations between humans – which is to say living beings – and technological organisms. The definitions are multifarious, and at the same time are intimately bound up with continuing discussions of the increasing amalgamation of human and technology. A widely prevalent characterisation of cyborgs refers to people who utilise bodily parts that have been supplemented through artificial components on a permanent basis (cf. Krützfeldt 2015). Anyone who wears a pacemaker or a cochlear implant can therefore already be regarded as a cyborg.

3   Currently, discussions of transhumanism are dominated by questions of the extensibility of human capacities through the utilisation of technical resources, as well as the associated ethical questions. At times, there are overlaps with the philosophical tendency known as 'posthumanism', according to which humanity is regarded as an obsolete evolutionary model destined to be supplanted in the near future by a new, successor stage of evolutionary development (cf. Kurthen 2011, 7–16).

4   According to Bruno Latour, *things* too can be protagonists to the extent that people act in conjunction with them in specific contexts of action (cf. Latour 2001). If disability is considered in relation to this model of actor-network theory (ANT), it becomes clear that designed *things* too have an active impact on the construct of disability.

5   Here, one might pose the basic question: what is the actual distinction between an assistive and a non-assistive technology? In this context, Sara Hendren is quite clear, stating that 'all devices are assistive' (Hendren 2013), by which she means that the majority of devices and objects intended for daily use are in all likelihood designed to support human beings in their everyday lives. Yet, according to Hendren, it is only in connection with the topic of disability that devices and objects acquired the peculiar connotation

of being 'special', i.e. of serving 'special needs' (ibid.). Accordingly, all 'devices' are at least potentially 'assistive' (i.e. a chair for sitting, a cup for drinking), and are however only designated as such in contexts where there is an explicit connection with disability, illness, or healing (i.e. a crutch for supporting some-one with restricted mobility).

6   With reference to the social welfare state model that arose in Sweden during the 1940s under the ae-gis of the term *Folkhemmet* (people's home), Malte Klein-Luyten describes the growing use during this period of 'democratic design' in the sense of a 'broad accessibility for all societal strata'. In its essen-tial aims, this coincides with the concept 'society for all', as proclaimed beginning in the 1960s by Olof Palme, Sweden's Prime Minister at the time. In Sweden, the topic of freedom from barriers and of the po-tential of design with regard to people with disabilities was addressed earlier and more extensively than in other countries (Klein-Luyten et al. 2009, 13).

7   The term 'inclusive design' is used in Europe and Japan, and 'universal design' primarily in the United States (cf. Pullin 2009, 2).

8   Here, the elasticity of the concept of design becomes clear, since even (or especially) political or legal tools as well can function as design resources.

9   Independently of the universal design context, the German professor of design Heiner Jacob (†), who consistently denied both of these claims in his seminars at the Köln International School of Design (KISD), coined the term 'Esperanto design'. He used it as a synonym for design attempts, which he re-garded as utopian, to produce objects that could be understood and used by all, and which in his view, in their striving to 'please everybody', must necessarily fail (Jacob 2004).

10   Sara Hendren points out that today, many 'assistive technologies' and devices are still conceptualised as 'medical aids', which often results – both materially as well as structurally – in a 'hospital aesthetic' which is unavoidably coupled (at least optically) with stigma: '"Assistive technologies" have largely taken their points of departure from medical aids, primarily because in industrialized cultures, people with atypical bodies and minds has been thought of as medical "cases", not as people with an expanded set of both capacities and needs. So a lot of the design attention to things like crutches, wheelchairs, hearing aids, and the like [has] followed the material look and structure of hospital gear. And accordingly, design-ers and people working in tech have "read" them as a branch of medical technologies' (Rosen and Hen-dren 2013).

11   In his theses on subjugating versus emancipatory design, Friedrich von Borries illuminates the degree to which even 'a putatively neutral [...] design, one supposedly devoted to the apolitical solving of prob-lems, cannot escape being bound up immanently with the political sphere'. For 'in many cases, a prob-lem-solving-oriented design stabilizes the existing order – and hence assumes a political function, albeit inadvertently' (von Borries 2016, 21).

12   Bazon Brock has stated that essentially, 'problems' can only be solved when accompanied by the emer-gence of new problems (cf. Brock 2011). This perspective links up with Horst Rittel's and Melvin Webber's consideration of so-called wicked problems (Rittel and Webber 1973).

13   Formal-aesthetic questions arise in particular when, for example, it is a question of designing a pros-thesis in a particularly discreet or realistic manner, so that it remains unnoticed or barely noticeable, or in contrast, whether it is especially conspicuous, is designed to avoid resembling the human body. Such decisions can be associated with divergent acceptance mechanisms: while eyeglasses evoke a variety of associations depending upon their appearance (wisdom, coolness, sportiness, etc.), and even function at times as a fashion accessory, the hearing aid remains – to the extent that it is recognisable – a stigma-tising factor. This raises the wider question of how we behave towards those whom we perceive as being afflicted by a putative defect.

14   Through the latest possibilities of miniaturisation and the use of transparent materials, the newest hear-ing aids by most manufacturers are designed in such a way that they are barely visible to others.

15   I.e. through virtually invisible contact lenses or hearing aids, through prostheses that resemble body parts or flesh-coloured orthoses.

16   In the design and innovation context, the notion of the 'target group' is regarded increasingly as a double-edged sword. On the one hand, it still seems indispensable to classify and define the intended users of a given design as precisely as possible. On the other, it appears difficult to exert any influence in advance on which users are going to use which products and in which ways. As demonstrated by Erlhoff and Brandes (Brandes and Erlhoff 2006), the concept of 'non-intentional design' clarifies the way in which, despite 'all of the competencies' of designers, design 'is actually only realized in use' (Brandes et al. 2009, 10). One of the great challenges for design, the result of industrial forms of production and associ-ated changes in markets, as described by Michael Erlhoff, is the circumstance that henceforth, 'the mar-

ket' has related to largely unspecifiable groups of people (Erlhoff 2013). In many cases, designers and marketing specialists attempt to cope with the resulting difficulties with the help of so-called 'persona', specific definitions of 'norm users', i.e. standard or average users. Problematic here, however, is the following: first, such standardisation practices often serve to cement dubious or false clichés concerning societal roles; and secondly, it is precisely those who already suffer from social marginalisation who are excluded by such categorisations.

17   The English term 'special needs' is a common albeit contested term for circumscribing individuals who require help or assistance.

18   Graham Pullin uses the term 'mainstream design' to characterise objects that were/are designed for a broad mass of people, or those which achieve success on the market, i.e. are sold in large quantities (cf. Pullin 2009, 89).

19   This involves certain difficulties. Craig Bremner refers to some of these, characterising the negative aspects of 'user-centred design', among other things with reference to terminological trickiness of the generally hazy term 'user' (Bremner 2008, 425 ff.). At the same time, Bremner draws our attention to the way in which a growing awareness of discrepancies between design and user intentions has led to a focus on new fields within usability research, i.e. participatory design and inclusive design: this involves attempts to understand the 'user' not, as in the past, exclusively as a 'passive recipient of predetermined signals […], but instead as an active and integral member of the design team', and at the same time to 'grant him greater participation than as the object of study and observation, which in the end confronts [him] only with something preexistent' (ibid., 427).

20   In her outline of a design methodology that is gender-aware and critical of power, Sandra Buchmüller shows how, from an economic perspective, human-centred design (HCD) is 'seamlessly compatible with the demands of mass production', but makes one thing clear, however: a preference for 'conventional and majoritarian oriented design solutions' necessarily stabilises existing 'relations of power and inequality in the respective field of application' (Buchmüller 2018, 258).

21   This distinction is formulated in a way that is oriented towards Margolin's classification regarding the politics of design, whose point of departure was found in Margolin's remarks on possible benchmarks for design and democracy (Margolin 2012), which were then reworked later by and with Ezio Manzini (Manzini 2017).

22   https://www.masterstudiodesign.ch/publications/integrative-design-design-research-approach-to-involvement/lorm-hand.

23   Design as *civic infrastructuring* deals with questions that relate to urban communities and their social practices, as well as to related possibilities and challenges of participation, civic engagement, community building, as well as questions of social sustainability (cf. Bergmann et al. 2013; Unteidig et al. 2013).

24   The intrinsically complex concept of equity should not be overextended at this point, and serves in particular to create a link here with the reflections of the sociologist of technology Werner Rammert, who understands equity and equality as essential values of social innovation (Rammert 2010). Meanwhile, the philosopher Richard David Precht calls attention to difficulties with this concept, which are traceable to the circumstance that there is no absolute equity, but only equity as subjectively perceived. With regard to its current social localisation, it is worth emphasising in particular the distinction between a liberal understanding of equity ('equity means that everyone enjoys the same opportunities') and a socialistic one ('equity means when everyone obtains an equal share'). According to Precht, all conceptions of equity in our societies represent attempts to negotiate between these two poles. The main problem is that ultimately, the criteria for defining equity and inequity are located differently by individuals and groups (Precht 2017).

References

Bergermann, U. (2013). 'Ability Trouble. Helen Kellers Handästhetiken'. In Bergermann, U. (ed.). *Disability Trouble – Ästhetik und Bildpolitik bei Helen Keller*, PoLYpeN. Berlin: B_Books, 15–54.

Bergmann, M., Herlo, B., Sametinger, F., Schubert, J., Unteidig, A. (2013). 'Community Infrastructuring – Designwerkzeuge zur partizipatorischen Stadtgestaltung'. In Lange, B., Prasenc, G., and Saico, H. (eds.). *Ortsentwürfe: Urbanität im 21. Jahrhundert*. Berlin: Jovis, 62–67.

Bieling, T. (2019). *Inklusion als Entwurf. Teilhabe orientierte Forschung über, für und durch Design*, BIRD – Board of International Research in Design. Basel: Birkhäuser.

Bieling, T., Joost, G. (2017). 'Hacking Normalcy – Disability from a Design Research Perspective'. In *Design for All*, vol. 12, no. 12, New Delhi, India.

Bieling, T., Joost, G. (2018a). 'Technologiegestaltung und Inklusion – Behinderung im Spannungsfeld von Technologie und Design'. In Burchardt, A., Uszkoreit, H. (eds.) *IT für soziale Inklusion: Digitalisierung – Künstliche Intelligenz – Zukunft für alle.* Berlin, Boston: De Gruyter,11–28.

Bieling, T., Joost, G. (2018b). 'Talk to the Hand! Digitale Inklusion von Taubblinden'. In Burchardt, A., Uszkoreit, H. (eds.) *IT für soziale Inklusion: Digitalisierung – Künstliche Intelligenz – Zukunft für alle.* Berlin, Boston: De Gruyter, 77–88.

Bieling, T., Martins, T., Joost, G. (2017). 'Interactive Inclusive – Designing Tools for Activism and Empowerment'. Book chapter in Kent, M., Ellis, K. *Disability and Social Media*. London: Taylor & Francis, Routledge.

Björgvinsson, E., Ehn, P., Hillgren, P.-A. (2010). 'Participatory Design and "Democratizing Innovation"'. In *Proceedings of the 11th Biennial Participatory Design Conference – PDC'10* (29 November 2010), Sydney, 41–50.

Borries, F. von (2016). *Weltentwerfen: Eine politische Designtheorie*. Berlin: Suhrkamp.

Brandes, U., Erlhoff, M. (2006). *Non Intentional Design*. Cologne: DAAB Press.

Brandes, U., Stich, S., Wender, M. (2009). *Design by Use: The Everyday Metamorphosis of Things*, Board of International Research in Design. Basel: Birkhäuser.

Bremner, C. (2008). 'Usability'. In Erlhoff, M., Marshall, T. *Begriffliche Perspektiven des Design*. Basel: Birkhäuser, 424–428.

Brock, B. (2011). 'Für einen neuen Umgang mit komplexen Problemen – Bazon Brock im Gespräch mit Stephan Karkowsky'. Deutschlandradio Kultur, broadcast on 7 December 2011.

Buchmüller, S. (2018). 'GESCHLECHT MACHT GESTALTUNG – GESTALTUNG MACHT GESCHLECHT. Der Entwurf einer machtkritischen und geschlechterinformierten Designmethodologie.' Berlin: Logos.

Dickel, S. (2011). *Enhancement-Utopien: Soziologische Analysen zur Konstruktion des Neuen Menschen*. Baden-Baden: Nomos – Wissenschafts- und Technikforschung,.

Dunne, A., Raby, F. (2014). *Speculative Everything – Design, Fiction, and Social Dreaming.* Cambridge, MA: MIT Press.

Ehn, P. (2013). 'Partizipation an Dingen des Designs'. In Mareis, C., Held, M., Joost, G. (eds.) *Wer gestaltet die Gestaltung? Praxis, Theorie und Geschichte des partizipatorischen Designs.* Bielefeld: Transcript, 79–104.

Erlandson, R. F. (2008). *Universal and Accessible Design for Products, Services and Proceses*, Boca Raton: CrC Press.

Erlhoff, M. (2013). 'Mit Michael Erlhoff über eine Theorie des Design', moderated by Jürgen Wiebicke, Philosophisches Radio, WDR 5, 12 July 2013.

Gesang, B. (2007). *Perfektionierung des Menschen.* Berlin: De Gruyter.

Hendren, S. (2013). 'All Technology is Assistive. Six Dispositions for Designers on Disability', Abler/Medium.

Hermes, G. (2006). 'Der Wissenschaftsansatz Disability Studies – Neue Erkenntnisgewinne über Behinderung?'. In Hermes, G., Rohrmann, E. (eds.) *Nichts über uns – ohne uns!: Disability Studies als neuer Ansatz emanzipatorischer und interdisziplinärer Forschung über Behinderung.* Neu-Ulm: AG Spak Bücher, 15–30.

Herwig, O. (2008). *Universal Design: Lösungen für einen barrierefreien Alltag.* Basel: Birkhäuser Verlag.

Imrie, R., Hall, P. (2001). *Inclusive Design: Designing and Developing Accessible Environments*. London: Routledge.

Jacob, H. (2004). Informations-Design, scholarly seminar, Köln International School of Design (KISD), Cologne.

Krützfeldt, A. (2015). *Wir sind Cyborgs: Wie uns die Technik unter die Haut geht.* Berlin: Blumenbar/Aufbau.

Kurthen, M. (2011). *Weisser und schwarzer Posthumanismus. Nach dem Bewusstsein und dem Unbewussten*, Hfg Forschung. Munich: Fink.

Latour, B. (2001). *Das Parlament der Dinge – Für eine politische Ökologie.* Frankfurt am Main: Suhrkamp.

Leidner, R. (2007). 'Design für Alle – Mehr als nur ein theoretisches Konzept'. In Föhl, P. S., Erdrich, S., John, H., Maaß, K. (eds.). *Das barrierefreie Museum – Theorie und Praxis einer besseren Zugänglichkeit.* Bielefeld: Transcript, 398–405.

Kercher, P. (2006). 'Design for All'. In *Design for All Institute of India* (ed.), newsletter (1) 2006; cited from Leidner 2007.

Klein-Luyten, M., Krauß, I., Meyer, S., Scheuer, M., Weller, B. (2009). *Impulse für Wirtschaftswachstum und Beschäftigung durch Orientierung von Unternehmen und Wirtschaftspolitik am Konzept Design für Alle*, expert opinion commissioned by the German Federal Ministry for Economic Affairs and Technology – BMWi, Berlin, 30 April 2009, IDZ Berlin.

Mace, R. (1985). 'Universal Design: Barrier Free Environments for Everyone', *Designers West* 33(1), West Hollywood / California, 147–152.

Manzini, E. (2017). 'The Politics of Everyday Life – How to Implement a Design-Based Collaborative Democracy', lecture, 20 February 2017, Carnegie Mellon University / CMU School of Design, Pittsburgh.

Margolin, V. (2012). 'Design and Democracy in a Troubled World', lecture, 11 April 2012, Carnegie Mellon University / CMU School of Design, Pittsburgh.

Mitrasinovic, M. (2008). 'Universal Design'. In Erlhoff, M., Marshall, T. *Perspectives on Design Terminology*. Basel: Birkhäuser.

Precht, R. D. (2017). 'Im Dialog. Gespräch mit Michael Hirz', *Phoenix*, 19 May 2017.

Pullin, G. (2009). *Design Meets Disability*. Cambridge, MA: MIT Press.

Rammert, W. (2010). 'Die Innovationen der Gesellschaft'. In Howaldt, J.,

Jacobsen, H. (eds.). *Soziale Innovationen – Auf dem Weg zu einem postindustriellen Innovationsparadigma*. Wiesbaden: VS Verlag für Sozialwissenschaften, 21–51.

Rittel, H., Webber, M. (1973). 'Dilemmas in a General Theory of Planning'. *Policy Sciences*, vol. 4, 155–169, Elsevier, Amsterdam; new edition in Cross, N. (ed.) (1984). *Developments in Design Methodology*. Chichester: J. Wiley & Sons, 135–144.

Rosen, R. J., Hendren, S. (2013). 'Why Are Glasses Perceived Differently Than Hearing Aids?', *The Atlantic*, 3 December 2013.

Sanders, Elizabeth B.-N. (2002), 'From User-Centered to Participatory Design Approaches', Frascara, J. (ed.), *Design and the Social Sciences*. Abingdon/UK: Taylor & Francis, 1–7.

Schöne-Seifert, B., Talbot, D. (eds.) (2009). *Enhancement: Die ethische Debatte*. Münster: Mentis.

Tervooren, A. (2002). 'Kritik an der Normalität. Disability Studies in Deutschland'. In *Das Parlament*, no. 29–30, 22729.

Unteidig, A., Sametinger, F., Schubert, J., Joost, G. (2013). 'Neighborhood Labs: Building Urban Communities through Civic Engagement'. In Proceedings of the Participatory Innovation Conference 2013, Lappeenranta University of Technology, Lahti, Finland.

DESIGNING THIN INTERFACES BETWEEN ARTIFICIALITY AND SOCIETY

Sandra Groll

Design for a Complex World

Design is an omnipresent phenomenon of our contemporary everyday culture. No matter if it's a lighting design, a sound design, an advertisement, a classic product design or the design of capital goods, hardly any artefact category will be spared the well-intentioned access of the design. All everyday life seems to go through the filter of professional design. It may be argued that this is nowhere near recognition with novelty, since the material endowment of the world has always been shaped by human effort to design one's environment. The result of that effort always had its effects on the form of social interaction in the society itself. The need to design everyday things and to give these things a social meaning through design is based on the trivial circumstance that the artefacts cannot shape themselves. However, they must be designed in a specific way to play a meaningful role in action contexts and social interactions in order to positively influence these interactions. The contingency problem of the appearance of the artificial co-societal society can only be solved by design.

The term 'contingency problem of the artificial' means that each artefact in its specific design is determined by a specific choice from an array of possibe options. Artefacts are thus always subject to an intentional design and thus to a previous observation with which the parameters of the resulting form are determined. The designed mode of appearance, being based on previous observations of society, bodies, and other artefacts, depends in turn on everyday practices, technological conditions, and the symbolic needs of the society.

Design itself is not yet a design in the modern sense. Rather, design is a fundamental cultural technique that owes its existence to the specific world relation of man and whose form is historically changeable. This world relation to Helmut Plessner (1975) can be understood as 'eccentric positionality' or, as Arnold Gehlen (2016/1940) suggests, as a consequence of the state of being of humanity, which is essentially determined by a lack of environmental adaptation and thus dependent on creating an artificial environment. However, the specific shape of everyday artefacts is more than the mere form of things, it is also a crystallisation of a whole range of cultural parameters and allows conclusions about technologies, social conditions, and social ideas. In this way, the design of the everyday could also be understood as a medium by means of which it is possible to communicate in a non-linguistic manner and that allows assumptions even when the actual symbolic and cultural codes have been forgotten.

Although the design of artefacts always fulfilled higher social functions, that is, communicating much more than the actual purpose of the respective object, it is necessary to make a clear distinction between pre-modern forms of design and contemporary design. To operate design under modern conditions means that every design has been preceded by a second-order and increasingly third-order observation, and that certain selections one observed in this observation have flowed into the design. The shift to second-order observation is due to the complexity of modern societies, and it may be said that the ongoing transition to a third-order observation is due to the emerging degree of complexity of a next society (Drucker, 2002).

While a first-order observer concentrates on the thing to be designed, that means he/she clarifies the what-question, a second-order observer observes observers, who in turn make observations. This form of observation is conscious of the fact, that the what of a design depends on how other observers look at it. A second-order observation can observe that an observer uses certain distinctions, which might differ from distinctions that another observer uses. The observation of an observer can, of course, be observed by an observer: a third-order observation, which then has to be considered to be only a first-order observation at the operational level. This observer can then see that other observers also make a second-order observation and see the observer using different distinctions to construct their world. This observer then can discover possible common points of reference.

On the level of the first order there would be a design that deals with the appearance of things, differentiating between form and function (Bolz 1999, 31), without being aware that this distinction is used. A second-order observer can see that the first-order observer differentiates between form and function, and that this could be handled differently. At this level, paradigms and programmes can then be formulated, which are able to give the designers an orientation for which approaches are possible to handle the difference of form and function in a way that satisfies the needs and demands of users. That these paradigms and programmes can and must be handled differently in order to be able to withstand the complexity and changeability of social everyday life is the knowledge and the observation of a third-order observer.

This observer can see that dealing with the difference between form and function always lies in a nested context of the plurality of possible interests and observers. At this level, it can also be seen that the shape of things and their design, forms an interface that orchestrates the plurality of perspectives by following an integrative design approach that is aware of the fact that interfaces must always be designed selectively, to make heterogeneities mutually compatible with each other. Roger Häußling (2016, 38) shows this with a simple example: an on and off switch allows humans to deal with a complex technology. The switch protects people from the complexity of the technology, but it also protects the technology from the complexity of humanity. This can only work if not all aspects of the respective complexity have to be taken into account.

The transition from pre-modern design to modern design takes place at the moment when the creative disciplines are changing over to a second-order observation. This change also takes place in other areas of society and is characteristic of the functionally differentiated society of modernity (Luhmann 1997a, 151 f.). In the field of design, this comes about on a structural level of the design activity in the course of the nineteenth century, but is only noticed by design theories in the course of the twentieth century. This change also became necessary because of the rising complexity of everyday life due to industrialisation, a separation of hand and brain work, anonymous mass markets and various technical innovations that brought forth a number of new types of artefacts. One can understand Louis Sullivan's (1999/1896) confrontation with the large office building as such a second-order observation. Sullivan observes that the established forms to deal with the difference of form and function and solve the contingency problem of architectural design, are also otherwise possible and that it is necessary to change them to meet the needs of new building types, new forms of work, and new forms of social organisation. For him and the designers at the transition to the twentieth century, the question arises with the help of which paradigms, methods, and approaches, the shape of new things can be determined. These lead, as it is well known, to a functionalism which seeks to secure the quality of form and to solve the contingency problem of design, through a universal objective aesthetic that is based on a rational handling of the difference between form and function in a context of use and production. With the criticism of functionalism in the 1960s, the increasing criticism of the aesthetics of goods (Haug 2009) and the subsequent postmodern approaches to design, the plurality, heterogeneity and thus the symbolic-communicative dimension of design are brought into focus. At the level of creative practice, approaches and experiments now emerge that test alternative ways of dealing with the guiding distinction of design. However, the 'anything goes' of postmodernism in the field of everyday design ultimately also increases the need for reflection in the creative practice, since potentially everything works, but not everything is equally desirable. Reflection theories such as the Offenbach approach to product language (Gros 1983; Steffen 2000) or Product Semantics (Krippendorff 2006) react to this development and offer methodological approaches for creative practice, with the help of which the social acceptance of design drafts already can be tested during the design process. These approaches also clearly show that the role perception of designers is shifting. Away from the genius creative subject towards an expert in *gestalt*-mediated organisation of plurality.

In this development first forms of third-order observation can be discovered and new approaches for a contemporary integrative design that understands the relationship of form and function as something that stands in different observation contexts. The concrete design, in this understanding, forms an interface between partially non-translatable perspectives. These considerations pave the way for a contemporary design that pursues integrative approaches and in so doing, makes a not inconsiderable contribution to social synthesis. This finding seems, at first, to overestimate the achievements of design and seems to stem from the omnipotent fantasies of a discipline that has always been suspected of being superficial and trivial. Also, the design is not in the privileged position of providing a 'General Problem Solver' (Newell, Shaw, and Simon 1959). But it is also a form of communication, and furthermore a form of non-verbal symbolic-based communication. In this function, design is very well observed by social and cultural studies (Baecker 2015; Baecker 2007, 254 f.; Bolz 1999; Moebius and Prinz 2012). For a theory of integrative design, communication aspects are as relevant as problem-solving aspects, due to the fact that there can be no socially relevant solution for complex problems without a form of symbolic-driven communication.

If in the present almost every offer of meaning and communication has gone through the filter of design, one may rightly assume that the evolutionary conditions for this development are to be found in the complexity and plurality of modern society. From this perspective, the social role of design in the implementation and the possibility of social synthesis of modern and 'next societies' is more fundamental and should extend well beyond achievements such as aesthetics of commercial goods or purely aesthetic practice. However, the question arises, with the help of which theoretical foundations and assumptions does the practice of design find its support and starting points? The following considerations should give a first orientation towards what assumptions a contemporary design understanding, its integrative tasks could understand. A first step would be already to understand design as something that is able to form relational networks between different actors through the gestalt of artefacts. Relational structures that are temporary and changeable but allow the construction of regulatory structures.

Contingency

For this purpose, it is useful to understand design as a form of coping with contingency and thus as a sometimes very successful attempt to reduce the risk of being disappointed under unclear temporal, factual, and social conditions. An integrative approach to design needs to deal in its theoretical foundations with topics like contingency, plurality, and insecurity, and its practice with the integration of heterogeneous actors, and this is only possible if one is able to deal reflexively with the difference between inclusion and exclusion at the level of the design process.

The term contingency in sociology describes the circumstance that everything is always possible in a different way and this means that the so-being or so-appearance of things is based on the constructions and distinctions of observers (Luhmann 1984, 159 f). Other observers might disagree. According to Niklas Luhmann, contingent is that which is neither necessary nor impossible, and contingency also constitutes an intrinsic value of modern society (Luhmann 1997a, 93). At the societal level, various forms have established themselves to deal with contingency, design brings possibilities with it. Scientific knowledge for example. Designers know that much of their professional practice is coping with the contingency of things and a lot of ignorance or lack of knowledge at different levels. However, they do not allow themselves to be frustrated by their own contingency experiences, but instead use them as an opportunity to explore ways of producing a coherent and consistent design in the interactive loopholes of the design process.

Design in its present form is above all contingency management. It owes itself to the fact that technological innovations lead to the expansion of artificial equipment of modern society, and the everyday reality of modern societies offers a multitude of alternative possibilities of experience and perspectives. All this leads to an increased need to make things, communications, and mindsets so that they can be experienced as consistent and coherent even under the condition of polycontexturality. In polycontextural societies, uncertainty about what it is all about is a potential permanent state. Since communications through language and communications via symbolic generalised communication media are too limited in their structure to pick up and proceed additional meaning, additional forms of communication are required, which can take up excess meaning and keep it communicable. While using language you can only say things one after the other, not all aspects at once. If you use a symbolic generalised communication media like money, its binary coding (pay/no pay) may secure the success of a communication, but this media is also limited. Design and its media form is able to pick up excess meaning. One thinks only about the design of banknotes, which is able to communicate not only the nominal value, but also the topics and identity of the respective currency area. The need to design is also due to the fact that the structural form of modern society, namely the differentiation of autonomous functional areas, leads to the everyday reality of modern societies containing a whole range of alternative ways of experiencing things and everywhere additional sense is provided. It makes a difference whether I stand behind a counter as a customer or as an employee and representative of a company. For example, the design of the counter is a way to mediate between these realities, for example, by protecting or enhancing the employee's body to give him/her and his/her chosen role more authority from the customer's perspective. From the perspective of the customer role, a counter is also an indication that the process of a certain interaction can be expected; for example, making a payment or receiving information. At the same time, design of the whole situation can extend over different levels and affect the light, sound, and atmospheric situations as well as the bodies of the respective employees. In this way, in addition to the expected service offers, it is also possible to communicate the identity and values of the respective company. Design, which not only looks at the counter as a product design, but also deals with the ultimate horizon of society as an integrative design, must keep an eye on these relations and make them central aspects of the design concept.

The shape of things communicates in many ways and it organises spaces of possibility in which, for example, it forms the bodies, suggests certain actions, and moves others far away. It can strategically stabilise or break conventions, but it always has an impact on social interactions and the possibilities of this interaction. The fragmentation of everyday reality into differently tangible realisations of reality considerably increases the complexity of everyday reality and leads to an increased need for forms of communication that are able to communicate additional meaning in a non-linguistic way. With the conversion of the dominant social structure to functional differentiation – individual functional areas assume certain functions for the overall system of society (see Luhmann 1984 and 1997a) and each develop their own perspective and special realities – a whole series is differentiated from new social roles. Within these roles too, different perspectives on everyday reality are establishing themselves. A simple example clarifies the problem and at the same time points to the functional area of an extended design with an integrative claim: a simple entity like the forest always means something else from the perspective of different actors.

From the perspective of the economy, the forest is a property and an investment that has to be protected from certain influences (such as the bark beetle). The interest here is in the quantity and quality of the wood and the price this wood can achieve in the market. Forest trails and their nature are only interesting if they allow, for example, modern harvesters to use them. The interests of riders, walkers, and mountain bikers play no role in this perspective or are perceived as disturbing. Depending on the field of research, forest is a different subject of knowledge for science. Nature conservation is interested in biodiversity and the preservation of biotopes. Bark beetle infestation is less of a problem, because it is a necessary intermediate step for the renewal of the natural forest. For the individual recreational athlete, the forest is a recreational offer that is subject to certain expectations: as natural as possible. But please do not be so natural that you suddenly face a group of wild boars. The list can be extended by various actors and other constructions of reality, even non-human entities could be found for which the forest simply represents the actual biological niche or habitat. It is also clear that the forest is an artefact that is produced in our speech and practice and that for different observers takes on different appearances and meanings. The field of application of an extended integrative design approach would be, for example, the design of a usage concept that corresponds to the different actors and their interests, takes place on different material and media levels, and thus conveys a relational structure between these different actors.

The everyday reality of modern society offers more opportunities for action and experience, and this means that experience of contingency – that is, the experience that everything is possible in principle – becomes more likely. However, if other options are available for every action and experience, then the risk of being frustrated by one's own reality-constructions, decisions, and expectations in contact with others also increases. Design is a way to deal with these contingent loads, but not to dissolve them. Any design is temporary and establishes a state that provides new opportunities for experiencing and becomes the subject of various references.

Contingency experiences do not only take place on the level of the individual experience, but extend to all areas of society and occur within all social institutions. In addition, functional differentiation not only leads to incongruent perspectives on the functional systems, but also favours the emergence of institutions and organisations, new social roles, and specific interaction situations for which the same applies. But even at the level of artificiality, this increases the potentially tangible complexity. Innovation, organisation, and new technologies are bringing forth a whole new set of artefacts that now populate what we commonly call everyday life.

As artefacts, they have a basic contingency problem: their form is not natural and has to prove itself in contexts that themselves are characterised by plurality and potential consistency of experience. They can appear in one way or another and still have to make sense for a wide variety of actors in their day-to-day activities and reality provinces. At this level, too, there is a triad of contingency, plurality, and uncertainty. With the differentiation of today's design, an autonomous field of observation and function has established itself with the plurality of everyday life and is busy translating the observed contingency into coherent and consistent material and media offerings. This succeeds design in which it empowers the medium of design to mediate relationships between various entities. Design is involved between physical systems, cognitive systems, social systems, and artificiality and tries to influence, the operations of those systems by gestalt in a desired manner.

Design as a Medium

First of all, it helps to understand the shape of things as forms of a medium (Heider 1926/2005). Each design organises selections, that means differences, that are fixed in certain material and medial substrates into a tangible form. The final form of a thing owes itself to crystallised individual decisions. To put it bluntly: this material, not that, this radius, not that etc.

At this point, I deliberately refrain from speaking exclusively about materiality, as the efforts of design are no longer limited to material things, but can encompass complete interleaving of different material substrata. It is important to note here, that the final shape or appearance of an artefact is due to a large number of individual distinctions, which include both a designation value (this) and a reflection value (everything else). The designation values can be sensually experienced, but not the reflection values. They can be reconstructed analytically, in which one recognises, for example, that a certain choice of material is motivated, for example, by certain technological conditions. Thus, the final shape of a chair can be analyzed into a series of individual distinctions: this seat height, because the seating must be suitable for human bodies that are accustomed to European seating conventions. This material, for example, because it seems to be best suited for this possible environment of use or in its effect corresponds to current social ideas, etc. Depending on the design task, the list can be supplemented with a number of aspects. It should be emphasised, however, that the final figure communicates as compact communication (Luhmann 2008, 146) about all these aspects.

The form and appearance of things forms a symbolic and non-verbal communication medium with which physical, cognitive, social, and increasingly also artificial systems can be achieved. As a proto-linguistic medium, the design of artefacts of everyday life is structurally comparable to the artwork in the art system (Luhmann 1997a; Luhmann 2008, 423). However, the design-driven design of everyday artefacts fulfils a different social function. In design, it cannot be a question of demonstrating the contingency of form (Luhmann) or communicatively keeping a fictitious or imaginary reality of society available. This is the social function of art and artefact 'artwork'. What design does instead is communicate consistency and coherence. And with that enables shared tangible real reality, even if one must expect the plurality of experience and different perspectives.

As a medium, the 'design' thus also enables the structural coupling of communication, individual consciousness, bodies, and artificial entities. In sociological systems theory, the term 'structural coupling' refers to the phenomenon that autopoietic systems, which have to make do with no operational contact with their environment, are at the same time dependent on an environmental relationship (Luhmann 1997a, 770). Since social systems cannot think and psychic systems cannot communicate, at this point the language allows mutual irritability. Similarly, it determines the shape of things, it is a more comprehensive sense of perception, body aligned, aesthetically and symbolically coded. In addition, the medium of design also makes use of the structural coupling between cognitive systems and society, in which it evokes specific meaning. Helmut Willke (2005, 15) points out that even language forms a symbolic system that presupposes *gestalt* perception and is evolutionarily dependent on the capacity for symbolic thinking.

In this way, the system-theoretical consideration of the term structural coupling can be described abstractly; it can be easily illustrated by a simple observation of everyday design: the lighting design of a room – its exertion has an influence on the form of communication, has an influence on whether, for example, certain roles in communication situations are accepted. Of course, it also has an influence on the mood and the thinking of individuals and thus has an influence on the form of the ongoing social interaction itself. However, the multifaceted benefits and omnipresence of design in contemporary society must not be misinterpreted as a determining social technology, as instrumental control or strategic control is limited by the plasticity of autopoietic systems. Since design is directed towards, however, the perceptible shape of entities and is basically serving the structural coupling between living, social systems, psychic systems, and artificial systems, but without being able to control them in their operations, the term 'contextual influence' might be useful at this point. With this term, possible omnipotent fantasies about the power of design would be rejected and causing a more modest conclusion to come to mind: design is a way of dealing with complexity and to control problems without being able to determine or want to determine.

Integrative Design

The historical development of design clearly shows that ever new aspects of context and observer perspectives are taken into account and integrated in the design process. This integration takes place with the help of very different basic assumptions, all of which, however, find suitable forms for dealing with contingency. This development is also reflected in the emergence of a reflective area of creative practice and thus in accompanying reflection theories.

Reflection theories are concomitants of the functional differentiation of society and serve the need for reflection of the respective practice as well as self-awareness (Luhmann 1997a, 965).

Already towards the end of the nineteenth century, initial reflections can be observed that focus on the relationship between design and society. For example, when William Morris (Morris 1890) speculates in his utopian novel on the possibility and appearance of a socialist society. Or when Louis Sullivan (1999/1896) reflects on principles for dealing with the new building type 'large office building'. And Gottfried Semper (1851) not only criticises the quality of German industrial products following the London Great Exhibition of 1851, but at the same time makes proposals for the systematisation of education and suggests influencing the general social discourse about the aesthetics of everyday products through the establishment of art and craft museums. One of the consequences of this development is that the relevance of the context is discovered in the design disciplines. It becomes clear that this is not just a context of use alone, but also that this context is interlaced in other contexts, such as production, economics, and society. The observation and analysis of these contexts provides variables that help determine possible parameters of an entity's design. They form a relational network between different actors, assumptions, and entities.

Schematically, this is roughly represented as illustrated in Fig. 1. Readers interested in systems theory will recall the schema of George Spencer Brown's calculus of form (1997/1969) and, to an even greater extent, the multifaceted testing and application of the calculus in Dirk Baecker's work (Baecker 2007). The details of the theoretical foundations of the calculus within the framework of systems theory cannot be covered on this point. I use the model of forms as a schematic representation to illustrate possible solutions to dealing with the contingency problem of

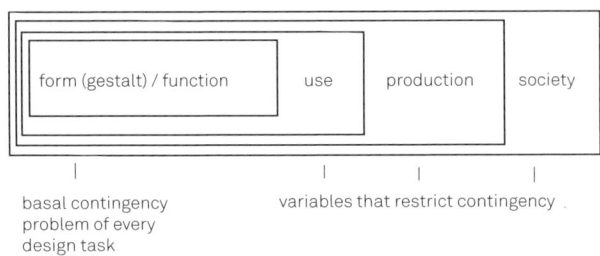

1 Design as observation: early state

| form (gestalt) / function | use | production | society |

basal contingency
problem of every
design task

variables that restrict contingency .

design. The schema above presents a combination of considerations of the form of design as a form of communication, as suggested by Baecker (2007, 265) following Norbert Bolz and the social model of form (Baecker 2007). It should be stressed at this point that every difference, every drawn-in separation in this schema, is also readable as an association. The difference between form and function can also be read as an association of a form with a function. This creates a state that can only be specified by the use of further differentiations or contexts. For example: use, production, and society. The schema can be read from right to left, as well as in the opposite direction. The schema presented above depicts the functionalist design understanding that dominated until about the middle of the twentieth century. Here, one makes use of the assumption that a generally universally acceptable form can be achieved if it is useful in the context of a utility function as well as in the context of industrial mass production. This approach is sufficient for a while to provide guidance and clarity on how to deal with the basic contingency problem of design, as it also correlates with central concepts of prevalent social semantics, emphasising rationality and objectivity. In this way, possible parameters can be identified that, as functional necessities, transform the indeterminate contingency of a design into determinate and create offerings that can be presumed to be experienced by users and producers as consistent and coherent. It soon becomes clear that design has to take into account more parameters in order to meet the complexity of its social mission. Alexander Mitscherlich (2008) criticising the inhospitality of the cities or Lucius Burckhardt's (2012/1983) objection that design has a non-visible side with impact on the form of social interactions, can be read as an indication of problems that tend to come to light under the paradigm of a functionalistic design that operates with a limited perspective of relevant contexts.

With the criticism of functionalism in the 1960s, there is a growing awareness that there are still other aspects and contexts that play a role in the designs of the design and must be taken into account. For example, the distinction value of a design in the context of groups, identities, and society. This development can also be expressed in the context of the schema.

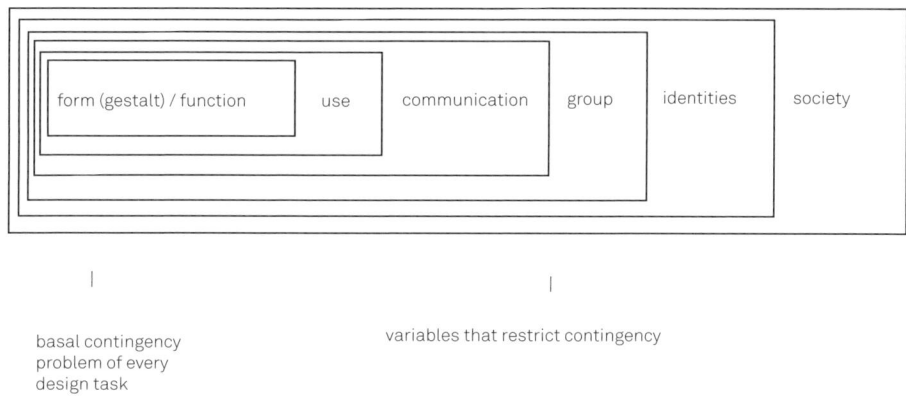

form (gestalt) / function   use   communication   group   identities   society

basal contingency
problem of every
design task

variables that restrict contingency

2 Design as observation: middle state

At this point, I have to shorten the historical outline of creative solutions to dealing with the contingency problem of design, and I would like to point out that in the disciplines commissioned with the design of everyday life, in the last 150 years, it is not just clearly a second-order observation, but increasingly explores the contextual framework in which a design possibly stands. This means that shape-defining decisions are increasingly no longer made on the basis of the genuine intuition of the designers, but rather as a function of a previous observation by observers and their everyday actions. This way of looking at things keeps capturing new artefact categories and at present, is no longer confined to simple material products. Klaus Krippendorff (2006) illustrates this in a trajectory of artificiality (Fig. 3).

At every level of this trajectory, there is not just an entity to be formed – whether it be products or goods, interfaces or multiuser systems, projects or discourses – but also the corresponding contingency phenomena based on certain observations (such as utility or intuitive usability) and issues (generativity and liability) must be translated into a coherent and consistent offer. A second-order observation, that is observation by observers and their ways of behaving, is central to all levels. In Krippendorff's trajectory, however, a circumstance which is of central importance to our current design tasks is not shown: the artificial categories are interlinked with each other. This means, for example, that products are also goods that correspond to the identity of a company and entail certain services. Each product forms an interface, at a simple level, between the body and the artefact, at a more complex level between the body, cognition, and the artificial system. Every design was and is a project, of the designers, of the companies, of the distribution, etc. Finally, every design is also part of a social discourse and a part of a discourse on design. And it is a component of a discourse about the form of our everyday reality itself and what one can expect from it. In this way, any design task can also be understood as a network. The design

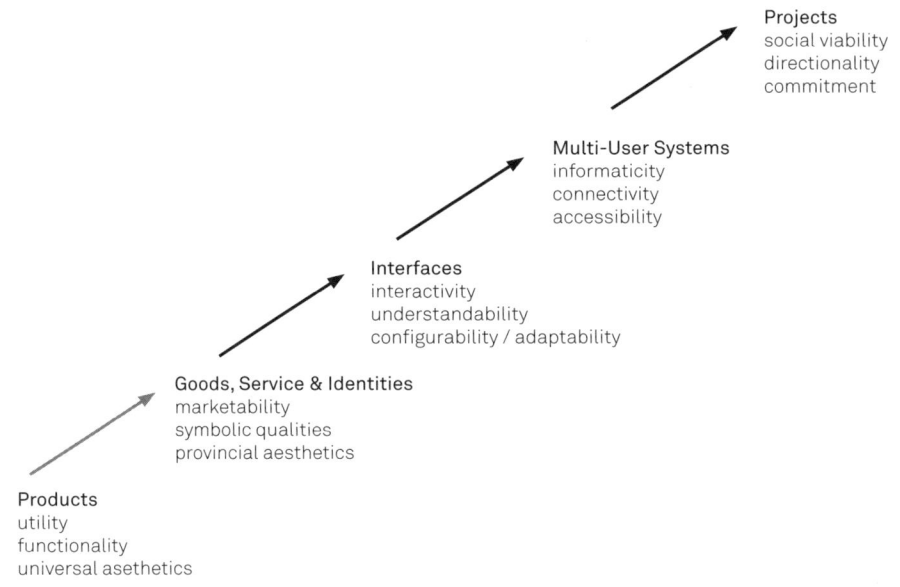

Projects
social viability
directionality
commitment

Multi-User Systems
informaticity
connectivity
accessibility

Interfaces
interactivity
understandability
configurability / adaptability

Goods, Service & Identities
marketability
symbolic qualities
provincial aesthetics

Products
utility
functionality
universal asethetics

3 Klaus Krippendorff, The Semantic Turn (2006)

of such entities must be capable of connecting with the interests, needs, practices, and constructions of reality of the most diverse actors. This task cannot be solved by further differentiating the design profession into other sub-disciplines, but requires a new generation of designers who are capable of solving complex relationships through integrative design approaches and working together with different expert cultures.

Dealing with Complexity

In design, a special handling of complexity seems possible, at least to suggest a number of approaches. Noteworthy here in addition are the considerations of Horst Rittel (2013) of the peculiarity of creative problems, namely their wickedness. And of course, Herbert A. Simon (1990/1981), whose science of the artificial could lay the foundation for a new generation of designers, should be mentioned. The systems theorist Ranulph Glanville (2011/2007) sees designers as scientists of complexity, whose special power is not only to deal with complexity on a daily basis, but also to possess methods and practices that help complexity to be treated as simplicity.

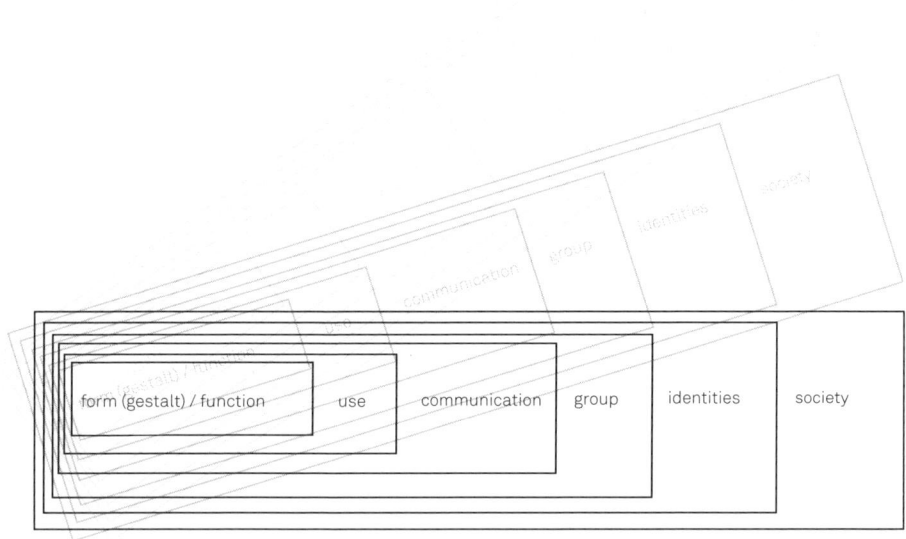

4 Dynamic model of design

From the perspective of sociological systems theory, Dirk Baecker (2015) even suggests that design could provide the crucial uncertainty absorption mechanism of the 'next society' and thus claim future significance comparable to that of myth and rituals in tribal societies, religion in ancient and medieval society, and the role of technology in functionally differentiated societies. As tempting as it seems to unanimously agree with such descriptions and understandings of design, the question must also be asked whether in the self-reflection of the design it has already developed suitable concepts, models, and approaches to guide the practice of design to fulfil its social function without having to rely on fortunate coincidence.

The signs increasingly seem to point to this. In addition to new curricular constitutions at the universities and significantly extended demands on designers in the labour market, there is also the burgeoning design science, whose brisk research activity is progressively accepted as an independent scientific discipline, not only as a sign of increased social claim on design.

The gestalt of things is not just a material form that can be so or otherwise, but rather a 'thin interface' between cognitive, physical, artificial, and social systems. Depending on the design concept and task, functional, material, or symbolic aspects can be weighted differently in the design of these interfaces. However, you will be able to identify these aspects in any design. They form a kind of relational structure through which the design of an entity is able to realise the structural coupling of consciousness, communication, and the artificial environment. In this way, the decisions made in the design have an influence on our everyday lives, our ways of dealing with them, structuring possibilities for action and communication. The question, however, is which models and assumptions the design practice is based on. What form and fundamentals must be adopted by a design that does not define its social task by means of the idea of generalised user needs and concomitant narrow functional context, but is aware of the integrative design of the insight that design ultimately always has an impact on the form of everyday reality. A design that is able to handle complexity and plurality, and is aware of the permeated heterogeneous constructions of reality, which are not just a question of a human-related design. Perhaps a dynamic model of design, in which the respective contexts are repeatedly traversed from the perspective of potential actors, can offer orientation here. However, in the practice of integrative design, it has yet to prove itself as an observation model.

References

Baecker, D. (2007). *Form und Formen der Kommunikation*. Frankfurt am Main: Suhrkamp.

Baecker, D. (2007). 'The Network Synthesis of Social Action 1: Towards a Sociological Theory of Next Society'. *Cybernetic & Human Knowing, 14*(4).

Baecker, D. (2015). 'Designvertrauen: Ungewissheitsabsorption in der nächsten Gesellschaft'. *Merkur* 69 (799), 89–97.

Bolz, N. (1999). 'Design als Sensemaking'. In Götz, M., *Der Tabasco-Effekt. Wirkung der Form, Formen der Wirkung*. Halle/Basel: Schwabe & Co. AG.

Bolz, N. (2001). *Weltkommunikation*. Munich: Wilhelm Fink Verlag.

Brown, G.S. (1997/1969). *Laws of Form*. Lübeck: Boheimer Verlag.

Burckhardt, L. (2012/1983). *Design ist unsichtbar. Entwurf, Gesellschaft & Pädagogik*. Berlin: Martin Schmitz Verlag.

Drucker, P.F. (2002). *Managing in the Next Society*. New York: St. Martin's Press.

Gehlen, A. (2016/1940). *Der Mensch. Seine Natur und seine Stellung in der Welt*. Frankfurt am Main: Klostermann.

Glanville, R. (2011/2007). 'Designing Complexity'. *Revue für postheroisches Management, 8*, 24–41.

Gros, J. (1983). *Grundlagen einer Theorie der Produktsprache* (vol. 1). Offenbach am Main: Hochschule für Gestaltung Offenbach.

Häußling, R. (2016). 'Zur Rolle von Entwürfen, Zeichnungen und Modellen im Konstruktionsprozess von Ingenieuren. Eine theoretische Skizze'. In Schimtz, T.H., Häußling, R., Mareis, C., Groninger, H. *Manifestationen im Entwurf. Design – Architektur – Ingenieurwesen*. Bielefeld: Transcript, 27–64.

Haug, W.F. (2009). 'Kritik der Warenästhetik'. In Haug, W.F. *Kritik der Warenästhetik. Gefolgt von Warenästhetik im High-Tech-Kapitalismus*. Frankfurt am Main: Suhrkamp.

Heider, F. (1926/2005). *Ding und Medium*. Berlin: Kadmos.

Krippendorff, K. (2006). *The Semantic Turn: A New Foundation for Design*. Taylor & Francis Group.

Luhmann, N. (1984). *Soziale Systeme*. Frankfurt am Main: Suhrkamp.

Luhmann, N. (1992). *Beobachtungen der Moderne*. Opladen: Westdeutscher Verlag.

Luhmann, N. (1997a). *Die Gesellschaft der Gesellschaft*. Frankfurt am Main: Suhrkamp.

Luhmann, N. (1997b). *Die Kunst der Gesellschaft*. Frankfurt am Main: Suhrkamp.

Luhmann, N. (2008). 'Die Autonomie der Kunst'. In Luhmann, N., Weber, N. (eds.). *Schriften zu Kunst und Literatur*. Frankfurt am Main: Suhrkamp, 416–427.

Luhmann, N. (2008). *Schriften zu Kunst und Literatur*. Frankfurt am Main: Suhrkamp.

Mitscherlich, A. (2008). *Die Unwirtlichkeit unserer Städte. Anstiftung zum Unfrieden*. Frankfurt am Main: Suhrkamp.

Moebius, S., Prinz, S. (2012). *Das Design der Gesellschaft*. Bielefeld: Transcript Verlag.

Morris, W. (1890). *News from Nowhere*.

Newell, A., Shaw, J., Simon, H. (1959). *Report on a General Problem-Solving Program. Proceedings of the International Conference on Information Processing*.

Plessner, H. (1975). *Die Stufen des Organischen und der Mensch: Einleitung in die philosophische Anthropologie*. Berlin: de Gruyter.

Rittel, H. W. (2013). *Thinking Design*. Basel: Birkhäuser.

Semper, G. (1851). 'Wissenschaft, Industrie und Kunst'. In Semper, G. *Wissenschaft, Industrie und Kunst und andere Schriften über Architektur, Kunsthandwerk und Kunstunterricht*. Mainz / Berlin: Florian Kupferberg, 27–71.

Simon, H. A. (1990/1981). *Die Wissenschaft vom Künstlichen*. Berlin: Kammerer & Unverzagt.

Steffen, D. (2000). *Design als Produktsprache. Der 'Offenbacher Ansatz' in Theorie und Praxis*. Frankfurt am Main: form Verlag.

Sullivan, L. H. (1999/1896). 'Das große Bürogebäude künstlerisch betrachtet'. In Fischer, V. and Hamilton, A. *Theorien der Gestaltung*. Frankfurt am Main: form Verlag, 142–146.

Willke, H. (2005). *Symbolische Systeme. Grundriss einer soziologischen Theorie*. Weilerswist: Velbrück Wissenschaft.

'DARK PHASES' AND 'DRAFT THINGS': ON THE INTEGRATION OF TECHNOLOGY AND DESIGN AS A CULTURAL PROCESS

Helge Oder

With reference to a concrete example, this essay demonstrates how design, within the context of a technological process of innovation, is capable of generating independent and significant results at various levels.

Investigated empirically in a number of praxis projects were approaches to design-driven product development with the aim of solidifying theses on the integration of design and technology. Dispositive for this approach is the intention of arguing for and making design comprehensible as a praxis-oriented knowledge process apart from technical contexts.

Introduction

Alongside decades of fulfilling practised, accepted disciplinary tasks, design is increasingly regarded as a means for reducing complexity, for thinking 'outside the box' in order to manage processes and collaborations, along with interdisciplinarity. Among the core competencies of design as well has been the development and shaping of identities (of trademarks and products, but in particular of lifestyles and their associated styles of consumption); in light of the necessity for post-growth strategies and less resource-intensive styles of consumption, the design of the architecture of decision-making, i.e. in the area of energy-saving measures, is also regarded as a form of design expertise. *Nudging* is the term used for such design strategies (Thaler and Sunstein 2008). Unfortunately, the design of the objects of these decisions often plays a subordinate role. Despite this, they are often the subject of overarching policy decisions, and often of technology-centred paradigms, i.e. in discussions of sustainability. Here, design often functions in tandem, assuming a mediating role by making technical developments comprehensible and implementable,[1] or in instances where technical innovation strategies encounter limits, to get the ball rolling (that is, once again, to think 'outside the box').

Cultures of Innovation

In this article, it is not a question of a stocktaking of the current background of design, but instead of an attempt to break down the relationship between design and technology and to sketch out future-oriented forms of this nexus – in other words, it is a question of competency for action. The *nature of design* that is characterised in the first introductory paragraph calls attention to a commonality: form and material are not accorded the status that is usually granted to them.

With reference to the example of mobility, transferable characteristics of innovation culture and technology can be adequately differentiated. Currently, the mobility of tomorrow preoccupies numerous actors from the business and research communities. Distinguishable in their approaches are two general models: first, there are the future labs where systematic approaches and large-scale mobility scenarios are developed for the medium and long-term. This approach involves explicit questions concerning urban planning, digital infrastructure, participation, and new business models. The originating competency of this design project is fuzzy with regard to the conflict situation, involving ideational, planning, and management competencies.

At the other end, industry is working concretely on new vehicles which, as prototypes and/or 'concept cars', are intended to demonstrate the potential (auto) mobility of tomorrow. In the second model, the fundamental implications and problem scenarios of automobility as a cultural and societal phenomenon are not adequately thematised. Questionable technologies such as the internal combustion engine are replaced under the auspices of *efficiency* and *consistency*. The role of design consists in the mediating of functions, in the design of interfaces and usability, as well as in profiling of products and trademark identities through styling.

The inadequacy of both models consists in the way in which the designed objects are handled. In long-term planning, these aspects are only addressed in a rudimentary fashion. Concepts that emerge as a result of planning are conveyed, or collaborative work serves the use of designed objects as containers for ephemeral meaning. *Design thinking* and the large-scale strategic and ideational aspects of planning and design demonstrate this quite clearly. Playing only a subordinate role is the design of more complex objects which would be developmentally viable with and usable by various stakeholders in ways that go beyond pre-established contexts of meaning. Only for specialised, delimited questions in the later phases of a process of innovation are such design approaches used for the sake of conspicuous differentiation.

Objectified with *technology-driven innovation,* again, is a highly specific vision of a proximate future. In the automobile sector, for example, high innovation costs ensure that innovation is restricted primarily to the level of technology and of the object, with maximal risk avoidance. Among other things, it is common to borrow technical innovation from other domains, for example networked digital technology, and to implement them in existing products (motor vehicles). Only afterwards does the production of serviceability, the design of interfaces, or the development of a periphery of more-or-less meaningful modes of use and user surfaces become a design concern. This design approach shifts objects and surfaces as carriers and mediators of the functional aspects of technology-centred processes of innovation into the foreground.

Technology-Driven Design

Complex technology, concrete procedures of navigation and data processing, as well as the technology that is found behind technical objects and structures, are difficult to understand, but – through well-designed interfaces – easy to use. On the whole, a prevalent ideal. To claim that the underlying technology is 'smart' in the human sense is, however, far off the mark. The powers of innovation and knowledge potential that are originary and intrinsic to design and to work on form are not exhausted by this 'superficial mentality'. The unavoidable consequence: design and innovation do not converge.

Useful in order to make this background more comprehensible is a brief historical excursion into the paradigm of design. Ever since the profession has existed, the anticipation of an appropriate solution to a pre-existing situation[2] has also been the task of product design. This approach, however, has its limits. Klaus Krippendorff (2013, 105) advocates the view that a connection must exist between a concrete design and the larger context of its interdependency, one that goes beyond its concrete purpose or function:

To design an artefact in such a way that it fulfils a specific function, [...] means to avoid questioning the larger whole that is to be served by the artefact and to reject any larger responsibility for a design (Krippendorff 2013, 350).

In functionalist social forms, what Krippendorff refers to as *technology-driven design* reinforces structures of authority and hierarchical inequality. As recipients of 'mass-produced products', argues Krippendorff, users are compelled to 'subordinate themselves, in the framework of functional subsystems, to the abstract understanding of the system as a whole' (Krippendorff 2013, 105). With this approach, it is only possible to do justice to the peculiarities and imperatives of real states of affairs and circumstances to a limited degree, and the design options of future ways of life are correspondingly restricted.[3]

Yet even dynamic, complex processes of change are based on structural foundations and framing conditions. To address these and contribute to their desired further development is – as Krippendorff demands – the domain of design: both in the more encompassing sense of *Gestaltung,* and of design project as a concrete undertaking (in the German language context, the *Entwurf*) (Krippendorf 2013).[4] This involves interactions between individuals and between individuals and the environment, as well as the interrelatedness between objects that are used within the spatiotemporal perspective in the sense of ecological systems. Under the premises of such an ecology,[5] Krippendorff understands the synergy between various artefacts (including products) which – assuming they are used – facilitate or displace one another, give rise to novel contexts, or cement existing relations. The automobile, for example, generates specific types of streets, traffic planning concepts, suburban settlements, but also technologies and value-creation chains, along with lifestyles and social roles. As a consequence, 'further' artefacts are shaped and engendered in a quasi-natural way. Krippendorff (2013, 256) speaks of a *technological cooperative* which is set in motion by concrete events, and which unfolds effects that are deeply staggered at the most diverse temporal levels, developing an 'ecology of the artefact'.

Technically driven and functionalist design paradigms, which are embodied both in the notions of functionalist societies as well as in the implicit, large-scale structures of the *technological cooperative*, are regarded by Krippendorff as inadequately linked to the larger context of design and its changeability.[6] In this context, Krippendorff opposes the notion of a *political heterarchy* against the notion of *functional hierarchies* which build upon one another: biological diversity and juxtaposition rather than goals, objectives, and authority. It is at this point that design comes actively into play. The correlation of societal forms of design culture, Krippendorff conjectures, may turn out to favour strongly non-functionalist and non-deterministic developmental and design concepts, which in turn promote and support pluralistic social forms.[7] In a nutshell: the object of design, ultimately, is always a form of collective life that fosters diversity and variation. Technical-functional relations are insufficient for shaping such contexts. But in contradistinction to Krippendorff, who sees the solution in the integration of stakeholders in particular in the ideation and planning, the present article sets the focus elsewhere: on autonomous design, and on work on form.

Sovereignty and Design

Examined more closely in place of technical contexts are the individually perceptible effects of technical and social complexity. The individual as the expert on his or her own lived reality, one directly affected by contexts of design and use, is at the centre of the most divergent design methods and design process models (i.e. human-centred design, participation, social design). The aim of such methods – alongside so-called 'actual needs' – is to make visible the characteristics and qualities of current and possible future ways of life. Peter Sloterdijk attributes to the individual the traits of an 'adaptable biomachine' which on the one hand can do 'less and less, but better and better', but which exists however in a 'psychosocial stimulus climate' which 'simultaneously provokes and annuls the sovereignty of the individual'.[8] According to this interpretation, we lack practices for the resonant linking of the lifeworld with the disposition of the individual in his capacities.[9] Sloterdijk infers from this that the individual must be offered a semblance of sovereignty. This simulated sovereignty concerns in particular the mastery of technical systems and objects. Design is accorded the function of securing the respective boundaries of competency, in the sense that 'placed in the hands of the subject are procedures and gestures which allow him to act as an expert in the ocean of his incompetence' (Sloterdijk 2010, 12).[10]

Sloterdijk marginalises the role of design and of the designer, regarded for the most part as unreflective actors who avoid complexity or any confrontation with complex factors of the lifeworld which are nonetheless susceptible to influence. Is the profession of (product) design unsuited to addressing complexity? According to this notion, the designer as the creator of the surfaces of black boxes, applies his skills to those few aspects that are comprehensible to him as an individual, and to attenuating the dissociation between directly comprehensible options for action and causalities.[11] With some justice, Sloterdijk compares this form of design to pre-modern rituals that serve as a means to master uncontrollable situations. Neither have any influence on the origins of the unease involved (Sloterdijk 2010, 12 f). I consider this strategy as inadequate if design is to have an impact beyond individual adjustment and sociotechnical training.

Jörg Petruschat too advocates the view that complexity is reduced by the process and performance of design and embodied in objects. In this, however, he does not perceive an act of concealment and of simulation, but instead of restructuring and of cognition. Petruschat (2011) writes:

In seeing and feeling the senses, we assemble a highly restricted selection of this data [note: which converge to the central nervous system from all zones of the body] in order to form a scene that allows us to act and make decisions. And this selection, this ordering, like an act of aesthetic pattern formation, also generates information on the basis of rejected complexity.

This way of framing complexity is, however, not a process of resignation. On the contrary, a quality is thereby created in the form of consciousness through which only the degeneration of volumes of data generates model spaces for action that go beyond embodied automatism [the initial capability of the instincts] (Petruschat 2011). This has far-reaching consequences for the creation of qualities in the form of mental models and material objects. In opposition to Sloterdijk, who grants the existence only of a design attitude that is restricted to surface activity and only simulates sovereignty, design for Petruschat rests on processes which reduce complexity at structural levels.[12] In this conceptual model, technology does not appear explicitly. Instead, it is a question of complexity that is nonetheless grounded to some extent – as described by Krippendorff – in technical-functionalist contexts and their social embeddedness. In contrast, the space of action of design is non-technical in nature.

Where Does Technics Come From?

To continue thinking through this aspect means to consider historical aspects of the quality of work on form in relation to technological innovation. The word 'technology' has its roots in the ancient Greek term *techné,* which can also mean art (or craft). But these concepts are not congruent. *Techné* refers to the abilities of the individual, to his skills.[13] Technique refers centrally to the enhancement and perfection of individual proficiencies in a resonant relationship with natural processes. Artistic competency can be built upon such a capacity for empathy. In the modern age, the term technology (and its counterpart technique) refers to a context of exploitation in which tools and resources are used to attain concrete ends (among them the enhancement of yields and efficiency) and the associated mitigation of functional (human) inadequacy. Petruschat (2003, 6 f) characterises the peculiar quality of 'techné-ical' processes as a form of knowledge of the effective repetition of natural processes in the human body:

While technology/technique is associated with a device, techné is bound up with the human individual, with action, experience, and self-knowledge. It is not merely knowledge of the repetition of natural processes, but also a subjective reflex, a construction, more precisely: a cognitive abstraction of specific invariant features which is established in gesture, in action.

With the concept of *techné,* human dispositions constitute the reference framework for success. It is only the reproduction of fundamental, invariant natural relationships through individual action and cognitive abstraction that constitutes the quality of this form of knowledge,[14] not imitative copying in conjunction with technical feasibility.

Like design practices themselves, the objects and paradigms of *Gestaltung* in the sense of sustainable forms of collective life have more in common with *techné* than with technology/technique – in the experience of the individual as well as in relation to the whole, in structures and contexts. The mental processes of model formation as well as their 'external storage' in the form of objects (drawings, pictorial objects, physical and digital models) represent one variety of a multiply determined form of action. It renders complexity operable in a non-technical way, generating knowledge as well as options for action on the level of the object, process, competency, and collaboration. Taking place alongside a multifarious 'feeling into' complex contexts is a continuous process of experience and self-knowledge in relation to the individual's non-technical capacities for reflection. The results are available in the form of various developmentally viable objects. The usage of the product is not a purely technical process of attaining a fixed aim. It must be regarded as an assessment of a proposal for a provisional approach to dealing with complexity. The concept of *techné* as a process of individual improvement forms the relevant background for evaluating the effectiveness of technology in the sense of apparatuses and devices. The human element is not stigmatised as deficient or improvable, but is instead the central object of the endeavour. Required as well in this context, self-evidently, are the technical contexts of function and purpose. They are not however paradigmatic.

Experiments and False Bottoms

A further perspective of technical contexts is offered by experimental procedures whose aim is a gain in knowledge.[15] Based on various studies of the practical realities of scientific practice in the natural sciences, Hans-Jörg Rheinberger has developed the concept of the experimental system. Remarkable from this perspective of scientificity is the realisation that not only knowledge itself, but scientific procedures as well are always characterised by vagueness and uncertainty. Rational, goal-oriented, reproducible procedures are seldom observable, yet these are often cited as the basis for the acquisition of knowledge – a manifest contradiction.

Secondly, the components of the experimental system are subdivided into vague, indistinct elements on the one hand and reliable, stable ones on the other. Rheinberger refers to the vague elements as 'epistemic',[16] and the reliable ones 'technical'. The latter category, it seems to me, was not conceived in relation to instrumentally rational paradigms of application, and instead in relation to the properties of things in a wholly concrete context of observation. This is also why Rheinberger attributes to these categories a potentially hybrid character. 'Technical' elements can be transformed into 'epistemic' ones and vice versa.

Whether or not 'technical elements' are reliable depends in particular upon the possibilities for action and application which they offer and open up.[17] This application does not proceed solely within systematically regulated procedures and functional options for viable action, as the nature of 'technology' would tend to suggest. Reliable elements can also have different origins. For example the result and objects of design activity. Their developmental viability, however, is not primarily technical in character. Initially, structuration and complexity reduction proceeds via a form of mental model formation. Factors that appear incontrovertible from a technical perspective can be variously weighted, abstracted, or ignored. The results of design activity are to begin with forms, which are accessible to perception and use beyond regulated, goal-directed contexts. They both alter and generate aims and functions. Decisive is the aspect of reliability, which emerges from a reciprocal relationship consisting of object characteristics, individual abilities, and context. Rheinberger refers to the fact that required of an experimenter, which is to say an individual who participates in carrying out an experiment, is a high degree of expertise. Only then is he or she able to set the experimental system into motion and to render the reciprocal relationship between technical and epistemic things productive for the sake of a gain in knowledge. All scientific results, Rheinberger concludes, are to begin with 'phenomena'. As a designer, one could also say: all objects are to begin with form. And the way in which a form is appropriated differs from the way in which an understanding of technical-functional contexts is acquired.

Alongside his microfactors, Rheinberger's approach can also be applied at a further level to the dualism between design and technology. From this perspective, the triad composed of the experimental system along with its 'epistemic things' and 'technical things' is also applicable to the objects of design. The framework within which each minimally responsible designer is consciously situated is a sustainable interaction between individuals, as well as a handling of resources, in short: of forms of culture. If this larger frame is regarded as an experimental system, then the object of cognitive striving, or the design respectively – the 'epistemic thing' – must be seen as a concrete form of interaction and cooperation between individuals. This should be improved continuously. Essentially, an endeavour that is associated more with *techné* than with technology. To remain for the moment with the analogy between the 'technical' and the 'reliable', the intermediate steps and results of development resemble 'technical things'. Among these are objects that are viable aesthetically for future development.[18] Usage, meanwhile, beyond simply perpetuating the already-known, counts as a process of knowledge production.

Development, Production, Use/Consumption

Considering these interim results now in relation to the aspect of the experimental system and of the shifting characteristics of *technical* and *epistemic elements*, we perceive a structural similarity between large-scale procedures and concrete experimental settings. There is no avoiding the fact that our life culture is shaped by the rhythm of technical procedures and structures. In many cases, as described above with reference to the concept of *techné,* a resonance relationship is rarely experienced by the individual – or only when value creation and resource use impinge indirectly upon stakeholders.[19] Processes of development, production, and consumption often engender contexts that are, as we know, neither desirable nor sustainable. Such structures are the primary focus of innovation and development. Implicit in every design, moreover, is a proposal which adopts a stance in relation to these questions. The object of design is a rapprochement between resources and individuals – ultimately, it is a process. Technology and clearly defined purposive and functional contexts are tools which form the basis for continuous experimentation. But qualitative developmental steps do not result solely from technically regulated procedures. The blurring of fixed role schemas in the triad formed by development, production, and use/consumption is a key component of design-driven innovation.

The Integration of Design and Technology

In the absence of concrete examples, admittedly, this theoretical background quickly comes to seem abstract. To begin with, it is a question of developing the imperatives and framing conditions of the technical contexts from out of the concrete design itself. A key role here is played by the above-mentioned developmentally viable objects. These objects engender novel modes of utilisation beyond technical contexts of purposiveness – through processes of innovation as well as through daily use.[20] Developed in relation to these stabilised, reliable, but nonetheless also provisional objects can be value creation chains, but also technological bases, production processes (structural classes for concrete objects), as well as framing conditions for digital production and the services and business models associated with them. Developmentally viable and 'embodied' in the objects, as well, of course, are approaches for new competencies and forms of collaboration within innovation processes. A concrete example can be seen currently in the area of e-mobility. Often, technical innovation takes place at the technological level in the absence of any concrete mobility concepts. As mentioned above, questionable technologies are replaced, efficiency enhanced, and consistency strategies implemented at the level of materials and fabrication. But how does one arrive at the concepts? Criticised at the start of this text were a variety of approaches to innovation; the example of e-mobility demonstrates that the deployment of diverse innovative products, in this instance e-bicycles, can lead towards the experimental exploration of new modes of use and cultures of mobility.[21] Certainly, such objects can include vehicles suitable for serial production. In their diversity, however, they have the character of prototypes in a larger experimental milieu. Possibly here is the manufacture and development of such vehicles through a variety of factors on the basis of rapid innovation cycles, and the accessibility of sophisticated and economical technical components. The implementation of electrified micromobility that is currently observable, which is to say everything beneath the level of the automobile (i.e. e-bikes), would not be possible without future-oriented investment by technology-oriented enterprises in the manufacture of such technical components.[22] Investigated via experimental paths to an increasing degree through the most varied types of usable objects are the most diverse forms of usage practices and mobility culture. In a second step, such new cultures – which are recognisable and describable in delimitation from the familiar, the habitual – can be grasped semantically and semiotically in an independent way, can be furnished with identities, and can become developmentally viable for direct participants.

Such an approach inaugurates innovative processes while at the same time integrating the lived realities of the most diverse stakeholders in design-driven and technology-based innovation – beyond ideative planning, abstract user surveys, or small-scale *UX* on the basis of prototypes.

Digital Production

Another thematic field where the aspects considered here are of relevance is digital production. In the future, we will be in a position to organise development, production, and use differently than hitherto. Novel forms of demand development and changing needs, local, decentralised fabrication and resource utilisation, as well as the displacement of developmental competence into the sphere of use and consumption are bringing significant new challenges into the design field. True here as well is something that was described above concerning experimental systems: the object of design is not primarily the usable object, but also the way in which various actors in diverse domains interact with one another. In this regard, impulses must arise from the design – and I would propose making them developmentally viable and relevant on all of the levels from development, production, and use/consumption. Digital production and Industry 4.0, conceived to date in a technology-centric way, have far from exhausted their full potential. Here, nevertheless, concrete technical questions, individual factors, and target/function constellations could be developed and resolved. Design could be one of the driving forces behind the integration of new technologies into our culture.

With reference to concrete projects, I have explored the way in which new framing and subtasks for processes of technical innovation can be developed out of a design. In collaboration with a research institute and a number of metalworking SMEs, it proved possible to develop a concept for a special, tool-poor form method of hydroforming. This procedural innovation brings together the themes of efficient, economical fabrication with the development of innovative products for independent solutions for existing needs, for example in the areas of container or plant construction (among other things the on-site manufacturing or the assembly of complex systems at inaccessible locations). Through highly individualisable objects, it becomes possible to rethink digital production and regional, decentralised value creation chains as well as the role of stakeholders. A precondition for this is to ensure that procedures are handled reliably on the purely technical level. The development of structural classes for specific requirements (stiffness, force transmission, etc.) must however be based on fundamental, non-technical impulses for these process technologies. Designed objects and the principles that are deducible from them form the reliable basis for the application of innovative methods of technology on the part of partners from business. Such parameters do not originate in marketing models or technical procedures, but instead design experiments and concrete work on form against the background of paradigmatic methods for establishing sustainable cultures of value creation. These stabilised, reliable objects are not primarily technical in character, although they are developmentally viable for technical approaches. Generated through experiments with form for participants with a background in technical socialisation is a space for critique and reflection. The objects cannot and need not be directly transferred into technical, systemically regulated fit locations in familiar procedures. They form the basis for evolving these habits and competencies through the proposals that are embodied in the design, without essentially negating existing knowledge cultures and forms of individual expertise. To refer to the results of such collaborations as 'technical' marginalises complex processes of design that are capable of conceptualising desirable futures.

Open Design – Closed Design

Design as a component of innovation and collaboration within enterprises has a long history. All of the German design schools, for example, whether in the pre-war era or in both East and West after the war, propagated participation in and attempts to influence production and value creation at an early stage. (For this reason, I do not regard 'design thinking', viewed in this way, as a new model, but instead as a fairly contemporary marketing strategy by IDEO.) There is, however, one striking difference from additional or historical approaches: earlier, the aim was among other things the development of good products that could be mass-produced. Today, this aim – a necessity that was at the same time generated by the history of industrialisation – has softened somewhat. It is primarily technical innovations in the domain of planning, communication, and manufacturing that are responsible for the fact that in particular more complex objects can be manufactured in small quantities or fabricated individually for manageable costs. This fact must be seen as the basis for the design of complex contexts and sustainable forms of economic activity and collective life. It is not simply a new framing condition for production. Instead, this circumstance contributes to breaking up the formerly technical frame for societal dispositions.[23] From the perspective of design, new paradigms need to be generated.

When we speak of the integration of various actors and the further development of competencies and roles, then we must also reflect on the developmental viability of designed objects for various stakeholders outside the concrete project setting or potential target group for recipients. Since developmental viability is not primarily technical in nature, we must assume that design results from earlier phases of a concrete developmental process can be developmentally viable and relevant beyond this process itself. In my view, this represents a variety of open design, one that should be deepened. More opportunities and risks may emerge from this for economically active players, since these processes and developments are for the most part accessible in their sub-steps, and hence divisible and communicable, their developmental viability having already been demonstrated. As the basis for the development of the individual as well as the collective beyond functional and sociotechnical training, in any event, this form of integrated design is worthy of consideration already today.

My personal experience with various collaborative projects gives a clear picture: for design activity *(Entwerfen)* to have an impact, it is necessary – throughout the various stages of a developmental process – to create spaces of autonomous design activity *(Entwerfen)* within a technology and planning-oriented and potentially open developmental process. Crucial for this are the 'dark phases' of the individual design. With the term 'draft things',[24] I have developed an independent concept that describes features and functions of artefacts in an epistemic-oriented design and development process. Further developing Rheinberger's approach to the 'epistemic thing', the aesthetic 'work on form' is emphasised as an independent process of cognition and unique design. These 'draft things' can be appropriated at an aesthetic level, and they embody knowledge and offers of collaboration. New meanings (semantics) and sign relationships (semiotics) can be made accessible to various stakeholders. Artefacts are not only experienced as containers for ideally prefabricated content or technical functional contexts. 'Draft things' represent qualitative and development steps that cannot be realised in another way, and which affirm the expertise and competence of professional stakeholders (for example in regional companies) and at the same time encourage their further development. Collaboration is thereby initiated and moderated. I would argue for a recognition of the particularity and autonomy of individualised design activity and of work on form as an indispensable form of praxis for the generation and concretisation of the desires of a sustainable living culture. Accordingly, the future role of the designer is that of a knowledgeable and reflective conductor of experiments. It is only their work that generates developmental spaces which make experiences possible and which are the objects of interdisciplinary understanding.

1 See among others Sloterdijk's 'Das Zeug zur Macht' (2010).
2 … or, depending upon the point of view, of a problem. In the framework of this work, the role of problem-solver which is often assigned to design is reflected on in a deepened way.
3 Precisely in open, non-limitable social dynamics, Horst Rittel acknowledged the difficulty of recognising the origins of problems and implementing ideas within a complex causal network. The familiar 'wicked

problems' are more relevant than ever before, and are central to design discourse. Rittel remarks: 'By now, we all beginning to realize thạt one of the most intractable problems is that of defining problems (of knowing what distinguishes an observed condition from a desired condition) and of locating problems (finding where in the complex causal network the trouble really lies). [...] As we seek to improve the effectiveness of action in pursuit of valued outcomes, as system boundaries get stretched, we become more sophisticated about the complex working of open societal systems, it becomes ever more difficult to make the planning idea operational' (Rittel and Webber 1973, 159).

4   Here, the concept of *Gestaltung,* – of design in the wider sense – is understood to mean the extension of life possibilities according to a self-model (according to Thomas Metzinger, a self-model is the experience of the affiliation of the body, emotions, and thoughts as a self, as well as perceptions of immediate familiarity with oneself). This concretely mediated form of identity formation rests upon cultural givens and also involves, among other things, the dissolution of an existing gestalt in efficacious elements and factors and their rearrangement in a new artefact. Design in the sense of *Entwurf* (which corresponds to the German verb *Entwerfen* = to design, project, conceive, conceptualise, draft etc.) refers to the linking of ephemeral ideas and concepts to a pragmatic basis, for example through pictorial-aesthetic experimentation and material work. See among others Petruschat's *Wicked Problems* (2011), retrieved from http://www.redesign.cc/Petruschat/Wicked_Problems_2.html.

5   This 'doctrine of the household' refers to more than the relationship between artificial and natural conditions.

6   With reference to the theme of the 'information society', Krippendorff describes and criticises a traditional conception of information that this preference for natural scientific knowledge and the 'know what' of factual information is 'in contrast to a "know how" which makes it possible to continually reconstruct and reshape worlds'. This concept of information described stable, and for various actors, conducive conditions which are nonetheless always temporary in nature, and require feedback coupling with changing framing conditions (Krippendorff 2013, 106).

7   An example of such heterarchical processes, according to Krippendorff (2013/2006, 351), is the development of the internet. It would be difficult to reduce its production to a planned process. Present instead is a temporally staggered process which links together various actors and their competencies, generating 'interim results' which for their part alter societal framing conditions, in turn contributing to the further development of the internet.

8   The individual, according to Sloterdijk, is nonetheless 'sovereign in his relative sphere of influence'. This position is, however, becoming 'more specialised and relativised' (Sloterdijk 2010, 11).

9   Here, the question must be posed whether the specialised capacities are not already or merely the result of sociotechnical drilling.

10 The possibilities of cooperation for the sake of reducing excessive demands, along with the resultant positive feedback experiences, are not the subject of Sloterdijk's reflections.

11 In this connection, Donald Norman brings in his interpretation of James Jerome Gibson's *affordance* approach and speaks of the *signifiers* (indicators of appropriate behaviour) and metaphors that must serve to explain interfaces. It should be noted here (as I remark later with reference to Petruschat) that the human body is the platform for all interactions; hence my approach to forming metaphors is based less on semantics and more on corporeal, pre-reflexive states. See Norman's *The Psychology of Everyday Things* (1988).

12 The reasons for this can be located to begin with in the similarity between individual perceptual and cognitive processes and design procedures. The formation of mental models does not simply follow functionalistic or technical-causal rules.

13 See among others Petruschat (2003), 'Befreit die Technik und ihr befreit die Form' in *form+zweck,* 20/2003.

14 See among others Petruschat (2003), p. 6 f.

15 I regard design activity as a knowledge-generating activity. See among others Donald A. Schön's concept of the *reflective practitioner*, as well as Jörg Petruschat's approach to design as a process of complexity reduction.

16 Meaning the object of the striving for knowledge.

17 Worth mentioning in this connection is the concept of *affordance*, developed by Gibson and made fruitful for design by Norman. The character of the *affordance* is intimately bound up with the dispositions of the individual. See among others Norman's *The Design of Everyday Things*.

18 I regard aesthetics as a dimension of the active appropriation and production in light of the biographical totality of the individual.

19 This includes processes which involve the exploitation of human and natural resources.
20 In light of rapid cycles of innovation and beta versions of products and services, it is in any event diffi-cult to separate development from use. Open and participative forms of development do the rest when it comes to blurring competencies and role attributions in these areas.
21 See among others Oder's 'Das Rad neu erfinden' (2017) in *Agenda Design 5 – Magazin der Allianz Deutscher Designer (AGD)*.
22 In this case, among others, the electric motors and guiding system for the e-bikes developed and pro-duced by the Bosch subsidiary eBike Systems. Here is an example of typical top-down innovation through which a relatively new context of use and development is opened further. The affordable avail-ability of technically complex, well-engineered propulsion components makes it possible for various ac-tors to realise the most diverse types of vehicles. See Oder (2017).
23 See Krippendorff's reflections on technology-driven design at the start of this text.
24 For the concept of 'draft things', see Oder's *'Entwerferische Dinge' – Things for the World* (2018). The Ger-man term 'Entwerferische Dinge' is not easy to translate into English. The literal translation of 'things of design' is insufficient to define the meaning of the term. Therefore, we want to use the term 'draft' in this text. Although it goes back to drawing, unlike my concept of 'entwerfen', it should suffice as a preliminary concept of work.

References

Krippendorff, K. (2013/2006). *Die semantische Wende*. Basel: Birkhäuser.

Norman, D. (1988). *The Psychology of Everyday Things.* New York: Basic Books.

Norman, D. (2013). *The Design of Everyday Things.* Philadelphia: Basic Books.

Oder, H. (2017). 'Das Rad neu erfinden'. *Agenda Design 5 – Magazin der Allianz Deutscher Designer (AGD)*.

Oder, H. (2018). *'"Entwerferische Dinge" – Things for the World'*. Lecture at FHNW Basel / Institut Integrative Design / Masterstudio.

Petruschat, J. (2011). *Wicked Problems,* retrieved from http://www.redesign.cc/Petruschat/Wicked_Problems_2_files/58_Petruschat_Wicked_Problems.pdf.

Petruschat, J. (2003). 'Befreit die Technik und ihr befreit die Form'. *form+zweck*, 20.

Rittel, H. W. J., Webber, M. M. (1973), 'Dilemmas in a General Theory of Planning'. *Policy Sciences*, 4/2, 155–169.

Sloterdijk, P. (2010). 'Das Zeug zur Macht'. In Sloterdijk, P., Voelker, S. (eds.). *Der Welt über die Straße helfen: Designstudien im Anschluss an eine philosophische Überlegung*. Paderborn: Wilhelm Fink Verlag, 7–25.

Thaler, R. H., Sunstein, C. R. (2008). *Nudge – Improving Decisions about Health, Wealth, and Happiness.* New Haven and London: Yale University Press.

AUTHORS

TOM BIELING is senior research fellow at the Design Research Lab of Berlin University of the Arts, where he heads the Social Innovation cluster. He has also been visiting professor at the University of Trento and the German University in Cairo (GUC) and is editor-in-chief at *DESIGNABILITIES – Design Research Journal*. In his research he mainly focuses on social and political dimensions of design. He has held numerous guest lectures and run workshops at international universities. He is a founding member of the Design Research Network, founder of the Institute for Applied Fantasies (Institut für angewandte Fantasie), and co-founder of the Civic Tech Lab at the Einstein Center Digital Future (ECDF). Falling Walls "Young Innovator of the Year" (2014). Recent books: *Design (&) Activism*; *Gender (&) Design;* and *Inklusion als Entwurf* to be published in 2019.

TONY FRY is adjunct professor, Creative Exchange Institute, University of Tasmania, visiting professor, University of Ibagué (Colombia), and Hong Kong Polytechnic University. Tony is a design theorist/philosopher/writer/practitioner with specific interest in design philosophy and politics, cities and post-conflict environments. He is the author of twelve books – his latest: *Remaking Cities,* London: Bloomsbury (2017).

SANDRA GROLL studied Design, Philosophy and Aesthetics at the Karlsruhe University of Arts and Design. From 2016 to 2018 she has been visiting professor for Theory and Practice of Design at the Kunsthochschule / University Kassel and lecturer for Design Theory and General Sciences at the University of Arts Bremen. She is a member of BIRD – Board of International Research in Design. Her research focuses on the social dimension and the role of design in modern and postmodern societies from a systems theoretical perspective.

ANNA MERONI is an architect with a PhD in Design, associate professor of Design in the Department of Design at the Politecnico di Milano. Her research focus is on service and strategic design for sustainability to foster social innovation, participation, and local development. A specific expertise has been developed in participatory design and co-design methods and tools. She is the head of the international Master of Science programme in Product Service System Design and coordinator of the POLI-MI-DESIS Lab, the Milan-based research laboratory of the DESIS – Design for Social Innovation and Sustainability Network. Anna is on the board of the PhD programme in Design, principal investigator of national and international research projects, chair of conferences, author of several publications, and visiting lecturer at international universities.

RALF MICHEL is a designer, design researcher, publicist and curator, and heads the research unit of the Institute Integrative Design of the Design and Art Academy HGK Basel. There he teaches design and research in the field of Integrative Design. In addition, he teaches Integrative Design in the field of Business Innovation at the University of St. Gallen. He represents a cross-disciplinary and humanistic design approach. Ralf initiated and directed the Swiss Design Network, initiated the Board of International Research in Design (BIRD) at Birkhäuser Verlag, and is still a member of the editorial board. He was a founding member of the German Society for Design Theory and Research. He curates exhibitions and is the editor of the book series *Schriften zur Gestaltung,* in which he publishes important texts from other languages translated into German. Ralf lives in Zurich and Sent.

HELGE ODER, a goldsmith by trade, received his Diploma in Product Design at the HTW Dresden, where he was deputy professor from 2014 to 2016. The main focus of his work lay in initiating research partnerships and conducting research projects between design institutes and SMEs. He taught as an artistic associate at the Bauhaus-Universität Weimar from 2010 to 2014 in the material and environment sectors. In his dissertation project at the BUW, he researches the autonomy of design in open design processes by means of experimental innovation projects. Within this context, Helge investigates the possibilities of early-stage optimisation of innovation projects by designing complex prototypes through integrative design. His findings are reflected in his teaching assignments, workshops, and publications. Helge lives and works in Berlin and Dresden.

JÖRG PETRUSCHAT is a design theorist. He studied Philosophical Aesthetics and Cultural Studies at the Humboldt-Universität Berlin. His dissertation in 1984 dealt with the functionalism of Shaker communities. He was editor-in-chief of *form+zweck* from 1985 and managing editor from 1991 (until 2008). In 1998 he became professor for Theory of Culture and Civilisation, and History of Design at the Design faculty in Dresden. Since 2004 he has been professor for Theory and History of Design at the Kunsthochschule Weissensee, with a focus on practice-based research. His latest publication is *Ungehorsam der Probleme,* Berlin (2017). Jörg lives and works in Berlin.

URSULA TISCHNER (BSc/MFA) studied Architecture, Art and Industrial Design in Germany. After being a researcher at Wuppertal Institute for Climate, Environment, and Energy, she founded econcept, Agency for Sustainable Design (www.econcept.org) specialising in eco- and sustainable design, and innovation of products, services, and systems. She has been active in training and education in Design for Sustainability with positions at Design Academy Eindhoven, Zurich University of the Arts, Savannah College of Art and Design, and is currently responsible for the Master Eco-Innovative Design of FH JOANNEUM, Graz, Austria. Ursula publishes books, organises conferences, networks around eco- and sustainable design, is a member of design juries and standardisation bodies such as ISO, and is an evaluator in European and national research programmes.

CAMERON TONKINWISE is professor of Design Studies and director of the Design Innovation Research Centre at the University of Technology Sydney. Previously, he was director of Doctoral Studies at Carnegie Mellon University, and associate dean Sustainability at Parsons The New School for Design. Cameron has a background in continental philosophy and continues to research what design practice can learn from material cultural studies and sociologies of technology. His primary area of research and teaching is sustainable design, and he is widely published on the ways in which service design can advance social sustainability by decoupling use and ownership – i.e., the 'Sharing Economy'. Cameron's current focus is transition design – design-enabled multi-level, multi-stage structural change towards more sustainable futures.

Translation from German into English:
Ian Pepper (texts Tom Bieling, Helge Oder, and Jörg Petruschat)
Project management: Nora Kempkens
Production: Bettina Chang
Layout and typesetting: Sven Schrape
Design concept BIRD: Christian Riis Ruggaber, Formal
Paper: Multi Offset, 110 g/m²
Lithographie: LVD Gesellschaft für Datenverarbeitung mbH, Berlin
Printing: Beltz Grafische Betriebe GmbH, Bad Langensalza

Library of Congress Control Number: 2019934226

Bibliographic information published by the German National Library
The German National Library lists this publication in the Deutsche Nationalbibliografie;
detailed bibliographic data are available on the Internet at http://dnb.dnb.de.

ISBN 978-3-03821-644-5

e-ISBN (PDF) 978-3-03821-531-8

© 2019 Birkhäuser Verlag GmbH, Basel
P.O. Box 44, 4009 Basel, Switzerland
Part of Walter de Gruyter GmbH, Berlin/Boston

9 8 7 6 5 4 3 2 1   www.birkhauser.com

BIRD

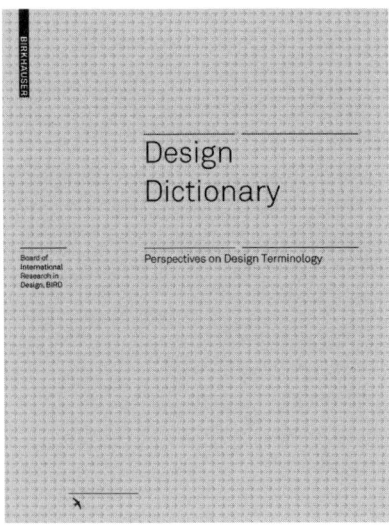

Design Dictionary
Perspectives on Design Terminology

This dictionary provides a stimulating and categorical foundation for a serious international discourse on design. It is a handbook for everyone concerned with design in career or education, who is interested in it, enjoys it, and wishes to understand it.

110 authors from Japan, Austria, England, Germany, Australia, Switzerland, the Netherlands, the United States, and elsewhere have written original articles for this design dictionary. Their cultural differences provide perspectives for a shared understanding of central design categories and communicating about design. The volume includes both the terms in use in current discussions, some of which are still relatively new, as well as classics of design discourse. A practical book, both scholarly and ideal for browsing and reading at leisure.

Michael Erlhoff, Tim Marshall (Eds.)
In collaboration with the Board of International Research in Design
472 pages
16,8 × 22,4 cm
Hardcover
ISBN: 978-3-7643-7738-0 German
ISBN: 978-3-7643-7739-7 English

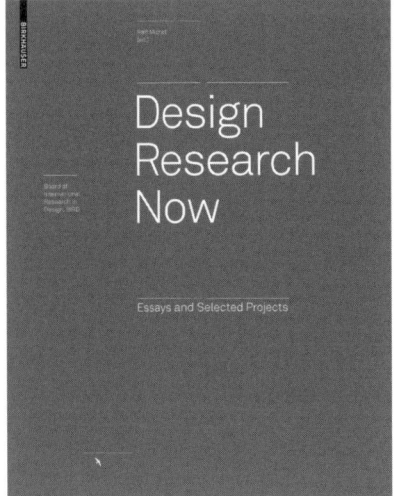

Design Research Now
Essays and Selected Projects

Design is becoming a recognised academic discipline, and design research is the driving force behind this transformation. *Design Research Now – Essays and Selected Projects* charts the field of design research with introductory essays and selected research projects. The authors of the essays, all leading international design scholars, stake out positions on the most important issues of design research. They locate the significance of design research at the interface with technological development, describe what makes it a necessary ingredient of the continued development of the design disciplines, and assign it a seminal role in the relevant developments of society.

The essays are supplemented by the presentation of recently completed research projects from universities in the Netherlands, the UK, and Italy.

Ralf Michel (Ed.)
In collaboration with the Board of International Research in Design
254 pages
22,0 × 28,0 cm
Hardcover
ISBN: 978-3-7643-8471-5 English

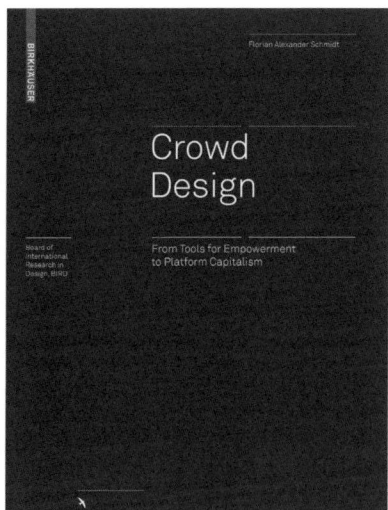

Crowd Design
From Tools for Empowerment to Platform Capitalism

The digital revolution is interwoven with the promise to empower the user. Yet, the rise of centralised, commercial platforms for crowdsourced work questions the validity of this narrative.

In *Crowd Design*, Florian Alexander Schmidt analyses the workings and the rhetoric of crowdsourced work platforms by comparing the way they address the masses today with historic notions of the crowd. The utopian concepts of early online collaboration are taken as a vantage point from which to view and critique current and, at times, dystopian applications of crowdsourced work. The study is focused on the crowdsourcing of design tasks, but these specific applications are used to examine the design of the more general mechanisms employed by the platform providers to motivate and control the crowds.

Florian Alexander Schmidt
In collaboration with the Board of International Research in Design
256 Seiten
16,8 × 22,4 cm
Hardcover
ISBN: 978-3-0356-1198-4 Englisch

Gender Design
Streifzüge zwischen Theorie und Empirie

Die Auseinandersetzung mit dem Geschlecht als sozialer Konstruktion ist in sehr vielen Wissenschaftsbereichen schon lange Teil der Theorie und Forschung. Im Design ist die Einbeziehung der Kategorie Gender allerdings noch immer ein blinder Fleck. Das ist merkwürdig, weil Design ja den ganz gewöhnlichen Alltag überall und jederzeit bestimmt und damit auch die in diesem Alltag handelnden unterschiedlichen Menschen. Und diese Interaktion zwischen den Subjekten und den Dingen findet unabdingbar „gendered" statt.

Das vorliegende Buch setzt sich erstmals mit den essentiellen Fragen von Gender im Design theoretisch wie praktisch auseinander: Es erörtert die grundsätzliche Notwendigkeit der Einbeziehung von Gender in den Designprozess, und es stellt exemplarisch Designprojekte zu diesem wichtigen Thema vor.

Uta Brandes
In Zusammenarbeit mit dem Board of International Research in Design
354 Seiten
16,8 × 22,4 cm
Gebunden
ISBN 978-3-0356-1227-1 Deutsch

BIRD

www.birkhauser.com

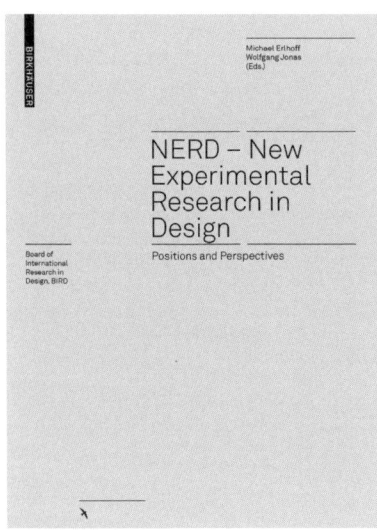

NERD – New Experimental Research in Design
Positions and Perspectives

Design has long expressed and established itself as an independent research competence – a fact that also companies, institutions and politicians have come to acknowledge. What is still needed, however, is a stronger public platform for design to confidently reflect upon this process and to establish and communicate the specific innovative and experimental dimension of design research. For this reason, the Board of International Research in Design (BIRD) has developed the New Experimental Research in Design (NERD) format.
The edited conference contributions of twelve young researchers from all over the world provide an impressive and diverse and insightful range of intelligent and inspiring approaches in design research, giving rise to further debate and action in the rapidly evolving field.

Michael Erlhoff, Wolfgang Jonas (Eds.)
In collaboration with the Board of International Research in Design
240 pages
16,8 × 22,4 cm
Hardcover
ISBN: 978-3-0356-1680-4 English

Inklusion als Entwurf
Teilhabeorientierte Forschung über, für und durch Design

Wie wir Dinge gestalten, hat einen maßgeblichen Einfluss darauf, was oder wen wir als „normal" oder „normabweichend" empfinden. Design markiert somit die Grenzbereiche zwischen In- und Exklusion, indem es implizit Rollen- und Wertebilder konfiguriert und dabei gleichermaßen in den Herstellungs- und Deutungsprozess von Normalität involviert ist. Wenn solche Normvorstellungen durch Design mitkonstruiert werden, bedeutet das im Umkehrschluss jedoch auch, dass sie sich durch Design dekonstruieren, also kritisch hinterfragen und verändern lassen: Design kann auch Gegenmodelle entwickeln.
Tom Bieling deckt auf zahlreichen Ebenen Verbindungen von Design und Inklusion auf und leitet daraus nicht nur neue Operationsbereiche für Designer ab, sondern liefert auch Anknüpfungspunkte für andere Praxis- und Wissensfelder.

Tom Bieling
In Zusammenarbeit mit dem Board of International Research in Design
320 Seiten
16,8 × 22,4 cm
Gebunden
ISBN: 978-3-0356-2020-7 Deutsch

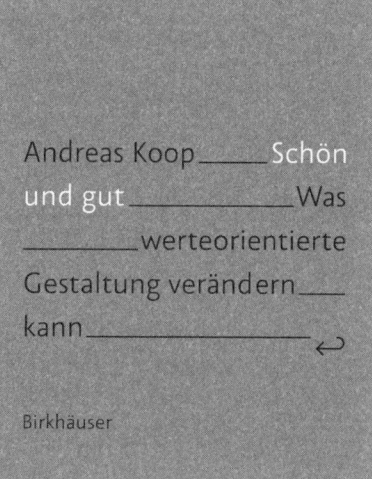

Schön und Gut

Was werteorientierte Gestaltung verändern kann

Was kann eine werteorientierte Gestaltung verändern? Indem sie Kommunikationsmedien, Produkte und Gebäude entwirft, die nicht nur einem ökonomischen Nutzen folgen, bringt sie zum Ausdruck, wie eine nachhaltige Entwicklung gestaltet werden kann.

Ziel des Buches ist, das Design aus der „Exekutive des Marketings" zu befreien und als Motor für akute, anstehende Herausforderungen zu begreifen: Sinn zu gestalten und Gestaltung Sinn zu geben, Kriterien für eine gute Gestaltung zu entwickeln und einen erweiterten Begriff von Design und Designforschung an der Schnittstelle zwischen Kunst und Wissenschaft vorzustellen. Damit Gestaltung ökonomischen, sozialen und ökologischen Anforderungen gerecht wird – und Design endlich die Rolle zukommt, die es einnehmen könnte.

Andreas Koop
160 Seiten
19,5 × 14,4 cm
Gebunden
ISBN: 978-3-0356-1829-7 Deutsch